GEORGE WAKEMAN

The Bones Beneath Us

Extinction, Evolution & The Fragile Age of Humans

First published by GW Press 2026

First edition

ISBN: 978-1-9193586-2-8

This book was professionally typeset on Reedsy.
Find out more at reedsy.com

Contents

Introduction

If Earth compressed its history into a single 24-hour day, humans would emerge in the few seconds before midnight.

A few seconds.

You've probably heard this before. It still makes my stomach churn when I do. Everything we call human history—from the first controlled flames and the cultivation of crops to the rise of ephemeral digital culture—would unfold within those few seconds.

It's an indication that our story, vast as it feels, is just the briefest chapter in a much older tale. For most of Earth's day, oceans, forests, and unimaginable creatures dominated the planet.

Despite this, we sit here today on artificial thrones, self-proclaimed apex species of our age. Animals living in rooms where we can alter the light and temperature with a switch. Scrolling through six-second simulated realities. Worrying about overflowing inboxes and saturated fats. We are primates in denial, apes in suits. We invented Microsoft Excel and the concept of Monday, and now suffer under both.

Extinction constructed the world as we know it. Continents split and collided, mountain chains submerged under rising seas, and ice sheets advanced and melted away. Sudden droughts erased species, shifting tectonic plates buried them, and toxic oceans caused them to disappear. Every living thing, including us, survives because its ancestors survived the end of a previous world.

We behave as if we have escaped it, as though nothing could halt the onward march of human progress. We build towers of steel and store our memories in "clouds". We assume tomorrow will always mirror today. But look back, and Earth's history tells a different story—of sudden resets, vanished rulers, and dominants who did not last long; others, however, endured for remarkably long periods.

To be clear, I believe credit is due. Regardless of our flaws, we are exceptional animals. If aliens landed tomorrow demanding answers, I could easily explain a hippo or a horse. But humans? That would take some work. No other species kills and dies for gods it invented in its own imagination. No other animal builds a global economy from numbers that don't exist, then binds billions of lives to it.

For all our absurdities, there is much to celebrate. We love intensely; we create art that outlasts the constraints of mortality. We contradict ourselves at every turn. Our existence is at once tragic, comic, and miraculous—which makes our story, and what came before it, even more surreal.

The Anthropocene—the proposed epoch defined by human-driven changes to Earth's climate, biodiversity, and geology—is where our story begins. Here, in 2026, the present day. Amid the buzz of smartphones, the turbulence of geopolitics, and the burden of a planet altered by our own presence. We begin here because, for the first time, one species has become a geological force, in a way no other life form has so far—reshaping the planet at a speed rivalled only by asteroids and volcanoes. From here, we will walk backwards through time, peeling back the layers of everything that came before.

You'll encounter animals that feel like they belong to fiction and relatives uncomfortably close. You'll pass through ages where animals ruled in forms you would not believe without fossil evidence, across extinctions and evolutionary stand-offs. Together, we'll trace the outline of humanity's story and contemplate the ghosts of hominids who almost were.

This book is an attempt at defamiliarisation, a repositioning, if you will. A way of seeing our existence as one improbable trajectory among countless lost alternatives. From that perspective, the modern world begins to look less like destiny and more like a peculiar glitch, an experiment gone haywire over the course of existence.

So let's head back on a path of rediscovery.

To remember what we are beneath the selfie filters, the name tags, and the uniforms. Eight billion upright animals, surviving day by day through cooperation, imagination, and borrowed time—one branch of life that has endured firestorms, extinctions, and oceans of time, now carving out a future that will, like every lineage before us, be folded back into the Earth.

How to Use This Book

Earth's history is difficult to imagine because the timescales are so vast. A single chapter may span hundreds of thousands—or even billions—of years.

To make that scale easier to follow, this book begins in the present and carries you backward through deep time. Through the rise of our own species, across the Ice Age landscapes that changed the world, beyond the long dominion of the dinosaurs, into coal forests where insects grew to the size of birds, to the shores where the first animals clambered onto land, down into seas where the earliest food webs took shape, and finally to the birth of the planet itself. Each chapter is another layer, like moving down through rock strata.

Once the present comes back into view, attention focuses on the human species. The final section examines our technologies and institutions within the same environmental and evolutionary history traced across earlier epochs.

The stratum diagram that follows works as your map. Earth's history is arranged as a series of stacked layers of time, just as rocks preserve the record of past worlds. Each layer marks a major geological boundary, and brackets indicate which chapter falls within that sequence. You'll see those same brackets repeated under each chapter title, giving a reliable point of reference. Dates are given in Ma (millions of years ago) and Ga (billions of years ago), the standard units used to mark time before the present.

If the scale ever feels overwhelming, return to the diagram. At a glance, it shows where you are in the long story of our planet. And if you want more detail, a glossary and a list of sources wait at the back of the book for further reading.

Timeline and Chapter Structure

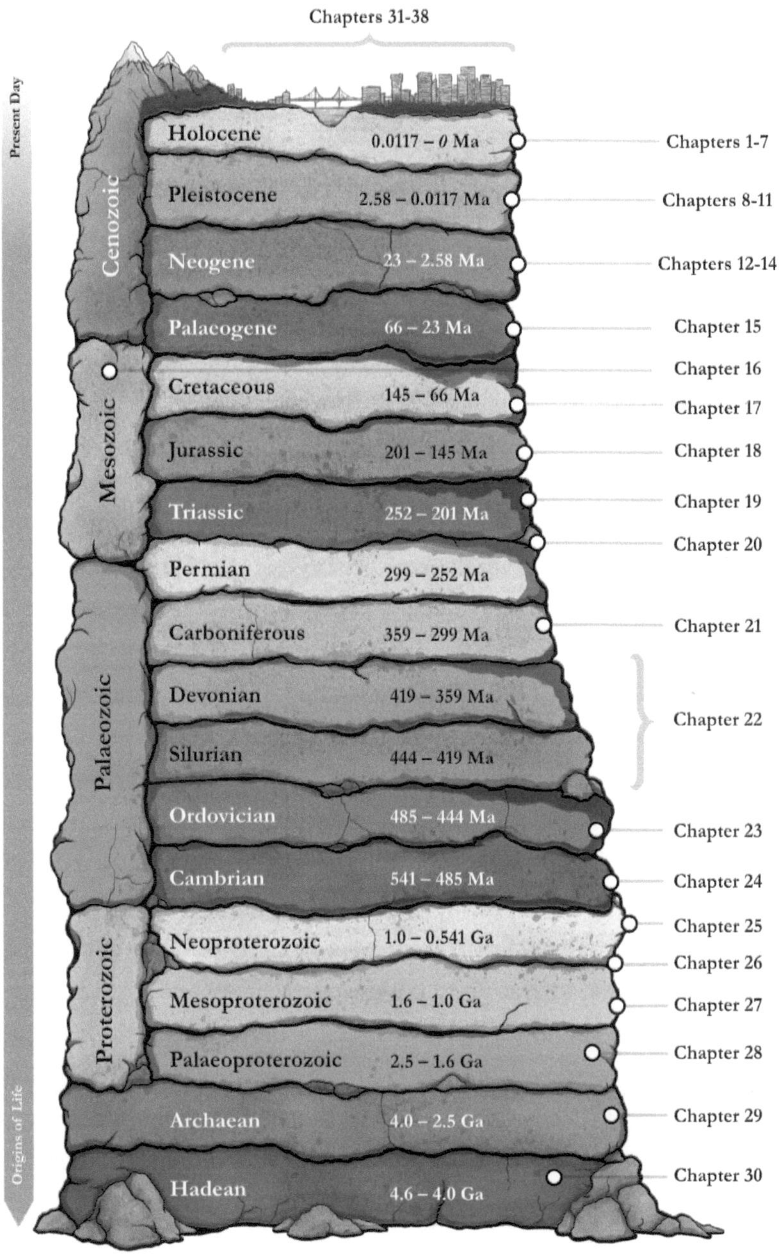

1

The Newcomers

Late Pleistocene–Holocene (≈300,000 years ago to present)

Homo sapiens are a biological exception: bipedal hominids with large brains, bare skin, and a taste for symbols, stories, and sugar. No other creature has changed the planet so rapidly. We have been here for just 300,000 years, yet in that short time, impacted nearly everything around us.

From a distance, nothing about us foreshadowed dominance. Chimpanzees remain much stronger than humans. Dogs and cats can outrun us. In brainpower tests, such as remembering numbers or opening food containers, many animals perform as well as, or better than, humans. If those skills were the sole test of intelligence, our rise would appear even more unlikely.

Our real advantage was different. We learned to build our lives around imagined things, such as gods, nations, currencies, and corporations. These inventions are not physical, but they shape our world as much as rivers and mountains. Shared beliefs became a force of nature. Ten silverback gorillas in a room fight for dominance, but ten thousand humans can organise into complex systems: bureaucracies, religions, space agencies, or even crypto bubbles. Large-scale cooperation became our strongest tool for survival.

Life spent millions of years slowly moving from water to land. By contrast, in just over twelve millennia, humans went from scattered foragers to city builders, empire makers, and digital network creators.

Agriculture started a chain reaction. As people stopped roaming and made the land productive, the land reshaped humans in return. Surplus food brought larger groups together. More people require more order. Order gave rise to hierarchy, taxes, disease, and war. Farming brought permanence: stone circles, villages, and monuments outlived their makers. These were the foundations of civilisation, fractured from the start.

Thanks to evolutionary chance, humans were ready for this shift. Other animals relied on instinct or brute force; humans depended on abstraction that reached beyond kin. Near the campfire, people mapped the heavens, traded stories, and passed knowledge down. Language became a toolkit; myth found meaning.

The human brain adapted to the subtle politics of the group. We tracked debts and loyalties, betrayals and alliances. Minds tuned themselves to status and gossip, instincts that still spark when a phone lights the dark. The circuitry that once bound us collectively also seeds rivalry and obsession.

Human bodies carry the signature of the savannah. An upright stance opened the horizon; sweat gave humans the stamina to outrun prey to collapse. Shoulders evolved as levers for hurling projectiles, eyes to follow their arc, tongues to shape thoughts into speech.

Today, most of us live far from the dangers our ancestors faced, and it is easy to forget how delicate life was for us until recently. Only in the last few centuries have tools and medicine stretched a net wide enough to hold populations of billions.

That net gives us freedom to imagine futures—and tempts us to believe they are secure; comfort suggests life will stay the same. Yet the stability humans enjoy is only temporary. Over the long span of existence, humanity remains an exception, and history shows that such irregularities rarely persist.

We are newcomers who have changed the world to suit ourselves. That change has given us empires, cities, and stories—and set the planet on a path it has never known before. The consequences of that legacy, and its costs, are what we will explore next.

2

How One Species Hijacked a Planet

Holocene–Anthropocene (≈11,700 years ago to present)

Describing Earth today to someone unfamiliar with it is difficult without reference to a particular species of primate. This species has transformed what was once a self-regulating planet into a tool for its own purposes. With a global population of eight billion, humanity's impact on the planet is undeniable. Yet the boundaries of our species' survival remain narrow.

From orbit, the planet glows with our presence. Coastlines glitter at night. Cities are stitched along the seam of land and sea. Cargo fleets traverse oceans in chains of light. Satellites hanging in space track these movements. The Amazon rainforest—once an unbroken canopy and long called the lungs of the planet—now wears the geometry of ranches and clearings. Squares and circles are embedded in its surface. The planet carries our signature as plainly as it bears tectonic scars or the courses of ancient rivers.

Technology now delivers every ecological upheaval as it happens. We watch the planet's crises unfold in real time. An alert, a clip, a breaking-news banner—each flashes across our screens almost instantly, only to be forgotten by the next headline. The information stream never pauses. Disasters arrive quickly; the shocks blur. We learn to live in constant upheaval. We observe until the damage to the biosphere itself becomes part of everyday life.

Peatlands that took thousands of years to form can disappear in a single season of fire. Glaciers whose melt once fed the great rivers of Asia and nourished millions retreat. Mangroves that softened the force of storms are stripped away for farmland.

A fisherman in Kiribati now ties his boat to the post his grandfather once used. But the post stands waist-deep in salt water. On the Tibetan Plateau, a shepherd watches pastures turn to swamp as permafrost collapses into thawing ponds. In Jakarta, a child wades to school through streets that flood more often every year. The sea claims land faster than concrete can resist. Across the planet, these lives are scattered, but each one forms a common thread in the same unravelling weave.

Meanwhile, the thawing ground unseals the bodies of Ice Age beasts. Their flesh surfaces from Siberian ice, defrosting from another epoch. Life attempts to adapt to our impact where possible. Oysters colonise the pillars of abandoned piers. Coyotes prowl Los Angeles streets as though the city were a fluorescent canyon. Fungi feed on radiation in Chernobyl. Insects redraw their ranges as the climate shifts. Life bends, finding margins where it can.

Even the air bears evidence of our presence. Jet streams stagger from their courses. Monsoons stall. Heat builds like a fever. Within that altered atmosphere, our traces remain: pesticide dust settles on Arctic snow, tyre particles in Alpine ice, microplastics lodge in bodies around the world.

The Hidden Weakness of a Connected World

Across the planet runs a mesh of human invention: fibre-optic cables on the seabed routing information, satellites carrying a planet's worth of signals, medicines and machines stretching lifespans beyond the wildest dreams of any king or pharaoh. From the bottom of the ocean to deep space, our actions now occur at scales that once belonged to the domain of gods. The long-lasting calm of the Holocene, with its 12 millennia of stable climate, was essential to everything.

Civilisations always trust that their world will stay put. When climate swerves or water fails, even the most intricate order can disintegrate. Our networks—dense, fast, and entangled—are far less forgiving than the life systems they run alongside. Nature survives by flexing. Spores drift after a cataclysm. Birds follow retreating ice. Mammals wait out extremes in burrows. We build speed and scale, but these very strengths become nodes of failure.

For now, civilisation cushions the shocks. Irrigation guards against drought. Markets guard against scarcity. Taxes keep firefighters and paramedics on call. But the same web that protects us entangles us. A hunter-gatherer band faces local risks of storms and famine. A modern metropolis is safer from day-to-day dangers, yet it depends on systems that will not bend much before they break.

Humanity as a Geological Force

While not officially recognised, many geologists argue that we have entered the Anthropocene—a layer of Earth's history defined by a single species. Concrete, plastic, aluminium, soot, and the radioactive trace of the twentieth century already document human activity in the rock. Whichever way one dates its beginning—early rice paddies and methane, the soot of industrial chimneys, the isotopes scattered by nuclear tests—it is already clear that the planet will remember us.

Each proposed starting point examines a particular leap: agriculture's first emissions, the engines of industry, the atom split into weaponry. Other labels to define it compete—Capitalocene, Plantationocene, Chthulucene—but none capture the whole. All concede the same fact: humanity has become a planetary agent altering climate, chemistry and evolution itself.

Zooming in, everyday life drowns this into abstraction. Markets flicker, data streams surge, screens brighten and fade while reefs bleach and insects vanish beneath the curve of our numbers. The signs we track—growth rates, stock tickers, GDP—mask the costs we inflict on the living world.

Our imprint is already permanent. A civilisation vast in reach, conscious of its own power, balanced on conditions that may falter without warning. A species that understands it can destabilise the Earth—and holds that knowledge before its collapse, not after.

Our control of the planet brings us to what many label a sixth mass extinction—a disturbance on the scale of asteroid collisions and volcanic and tectonic upheavals that have altered existence before.

3

The Sixth Mass Extinction

Past five mass extinctions (444–66 Ma) and today's Holocene–Anthropocene

The constant risk of extinction drives life forward. Species rise, flourish, and vanish. Their traces are sealed in time for us to discover or erased without a record. Earth's history is written as much through disappearance as through persistence.

Evolution continues in the gaps left behind by extinction, weaving fresh forms into old spaces. This is the background rhythm of life, steadily producing new life forms across millions of years.

At intervals, something threw the rhythm off balance and into upheaval. The Earth preserves moments when disappearance swept across the planet in waves. Entire dynasties ended together. These collapses cleared the ground for huge evolutionary pivots and unexpected beginnings. They came rarely, separated by tens or hundreds of millions of years. But their destruction modified the planet in ways no other force could.

Now we stand inside what many scientists argue is gaining momentum to become the sixth mass extinction. Its scale is nearly the same as the others', but its tempo differs. Except for the asteroid that struck Earth and killed the dinosaurs, the extinctions of prehistory unfolded over thousands or even millions of years. This one has gathered momentum over centuries. Estimates show current extinction rates are 100 to 1,000 times higher than the background rate of one to five species lost per year recorded in the fossil record. The force driving this acceleration is a single species. Chainsaws, trawlers, highways, and the promise of progress have become the new drivers of extinction.

For most of our history, we did not notice. Extinction advanced at a pace too slow for perception, or we were otherwise oblivious to the concept. Generations of humans lived and died without knowing that species were slipping away. We only know of the first losses that correlate with our rise—mammoths, moas, giant ground sloths—in hindsight. By the dawn of agriculture, humans had already reshaped entire regions. The old web of life had to bend around their expansion.

The pace accelerated with industrialisation. Coal furnaces blackened the air, mines opened the earth, and railways spread throughout continents. Species that had thrived abundantly for millions of years faltered within decades. The twentieth century carried this acceleration further. Climate change emerged as an additional pressure and a topic of contention.

Archiving Extinction

This extinction event is being documented in real time. We pin the last individuals in drawers, cryopreserve their tissues in freezers, and broadcast their decline in documentaries narrated in elegiac tones.

Guards provided round-the-clock security for the last male northern white rhinoceros. He died beneath floodlights, his image streamed across the world as if he were a dying head of state. Kakapo parrots in New Zealand breed under infrared cameras, with every single egg monitored. Scientists fit condors with satellite tags and monitor their movements, as we do with criminals. These are animals elevated to global fame only at the edge of their disappearance.

Zoos have come a long way in recent years. Most are modern bastions leading the charge on species preservation. But despite their work, they also show signs of a bigger issue at play. Zoos now serve as repositories of genetic memory. Species often survive here as ambassadors from habitats that no longer exist. We call them conservation centres, but often they are mausoleums. They hold on to lineages that cannot persist in the wild, stabilisers of native populations that are constantly fighting an uphill battle.

Museums hold the strangest records of all. Glass cases contain the skins of birds that once gathered in the millions. Now, only a handful of specimens remain. In the freezers of research centres, we preserve strands of DNA in liquid nitrogen. These banks are called "frozen arks," with the implication of rescue and future resailing. They are intended to act as insurance policies against finality. We preserve genetic fragments of what we continue to erase, as if foresight may compensate for action.

And in culture, we curate what we destroy. Our intentions may be sincere, but the irony is hard to ignore. Images of vanished species are printed on banknotes, carved into coins, and plastered onto beer cans as if they were mascots for sports teams. We place the dodo on postage stamps while birds alive today disappear. We teach children to recite the names of endangered animals. Then we sell them plush pandas and tigers whose wild populations are shrinking each year.

It is destruction paired with memorialisation, erasure paired with commemoration. We are the first species to cause such an event with our presence. We are also the first to be fully conscious of it. We mourn the disappearance even as we accelerate it. We write treaties, create protected areas, and fund conservation charities. Conversely, the drivers of human progress will always continue to take priority.

The Human Face of Mass Extinction

Extinction is, for the first time, both a biological and a cultural event. When a species disappears, it takes with it a set of meanings, rituals, and associations. Polynesian bird names fall silent as the birds vanish. Inuit words for kinds of ice lose their relevance as the Arctic warms. Australian Dreaming stories attached to animals and landscapes weaken as those species decline.

A child born today might age in a world devoid of familiar birdsong, living coral, and freely roaming elephants. The extinction crisis is also a crisis of the human mind. It is a narrowing of what the mind can picture as normal and what it can hope to inherit.

It is strange enough to instigate a mass extinction. Stranger still, to be its archivist, its brand manager, and its grief counsellor. We foresee our actions and measure them in real time. We mourn their loss beforehand. No species holds such power along with such weakness. No other species has complete knowledge coupled with such inaction. Only we are absurd enough to hold candlelight vigils for creatures whose habitats we bulldozed. We mourn in one hand what we profit from in the other.

This belief tempts us. It makes us exceptional and casts us as singular among life's lineages. But this knowledge changes nothing about the outcome. Knowing them does not slow down the rate of loss. The carbon in the air does not diminish because we measure it in parts per million.

If we are inside the sixth mass extinction, it feels like the end of the natural world. But the Earth has broken many times before. The planet has existed for four billion years and has ended worlds for most of that time. To our knowledge, extinction has occurred five times on a scale considered a mass extinction. Ours is not the first, only the first to be televised.

Palaeontologists use a hard threshold for a true mass extinction: three-quarters of species lost in a brief window. We have not reached that number. But the warning signs match the opening moves of past planetary die-offs. Coral reefs bleach and crumble; amphibians vanish from intact forests. Entire food webs thin long before their losses can be traced. Some of these losses include dates, photographs, and eyewitness accounts. Their demise still sits within family histories, field notes, and museum drawers. They belong to history we can still touch—a litany of extinction that has happened within the reach of memory.

4

Extinctions in Recent Memory

Miocene origins to modern extinctions (≈23 million years ago to present)

Most of Earth's creatures were lost before humans existed. Their shapes, biology, and habits faded into the sediment, leaving an absence recorded only on the slow timescale of geology. Yet, a few disappeared within reach of memory—preserved in paintings and photographs, labelled in museum drawers, and captured on early film.

Animals lost in recent history left evidence we can still access. We know their faces, and even some of the last survivors by name. Their images sit in glass cases, their remains embalmed, their likenesses painted or fixed on glass plates. These extinctions are not the most recent—many have followed—but they stand out for occurring under modern scrutiny. Telegraph wires spread news, newspapers printed rumours, and collectors sought rare specimens for cabinets—status symbols earned in blood.

Industries and markets, indifferent to decline, erased some species. Others fell to paranoia, condemned by human belief. Some existed in the wrong place at the wrong time—when humans arrived. While these forces ended individual species, the larger patterns—driven by illusions, appetites, and myths—continue to this day. What follows are stories of ancient animals confronting modern pressures.

A River Bereft of its Goddess

For millennia, the Baiji, or Yangtze river dolphin, swam through the silted waters of central China. Fishermen once spoke of it as the river goddess. Its lineage reached back over twenty million years, long before the first Chinese dynasties rose along the river's banks.

Within a century, the Yangtze transformed from a corridor for dolphins to an industrial artery filled with barges and dredgers. Concrete walls enclosed its banks, and chemical runoff tainted its waters. Propellers sliced through channels where dolphins had once hunted by sound. As echolocation met a wall of engine noise, gill nets and set lines turned side channels into hidden traps.

By the 1980s, only scattered animals remained. A 1997 survey recorded mere handfuls. During the winter of 2006, two research ships patrolled the river for six weeks, using observers and hydrophones to listen for the dolphins' calls. They detected none. The expedition declared the Baiji extinct—a dolphin that survived ice ages could not withstand a century of cumulative human pressures.

From Billions to None

Passenger pigeons once filled North American skies, their flocks so dense that they blocked daylight. Roosts stretched for miles, the weight of so many bodies snapping mature branches. The sound of their passing was a roar, as though the sky itself were alive.

Abundance seemed stable, and few doubted the birds' survival. Nothing so numerous, it was thought, could vanish. They seemed inseparable from the sky itself. Yet, industrial forces and organised killing challenged this stability. By the mid-nineteenth century, hunting became a coordinated enterprise. Telegraph operators relayed sightings of gigantic flocks; hunters converged with shotguns and nets; railways moved tons of carcasses to city markets hungry for cheap meat. In a short span, billions of pigeons were turned into freight cars of flesh.

Their absence reshaped the land as surely as their presence previously had shaped it. Evidence suggests that eastern woodlands that evolved under the pressure of roosts grew denser and more diverse once the flocks were gone. Those who observed the darkening sky for hours could not envision the spectacle ceasing until it was too late.

In 1901, someone reported that they shot the last wild pigeon in Ohio. Three decades later, only one captive remained: Martha, a solitary female at the Cincinnati Zoo. Visitors came to glimpse the final piece of a vanished sky while she watched them in turn. On 1 September 1914, she died. Her body was sent to the Smithsonian for preservation. The most abundant bird on Earth collapsed into a single specimen, then none. Ironically, their multitude proved to be their weakness: larger flocks were easier to track—and more profitable to destroy.

Counting Tiger Bounties

The thylacine, or Tasmanian tiger, was a striped, dog-like marsupial predator that once paced the forests of Australia and Tasmania, hunting and living in solitude. To us today, it is a symbol of extinction; to European settlers, it was a menace.

Suspicion surrounding the animal quickly cemented into policy. In the early 1800s, colonial governments offered bounties for carcasses. Between 1888 and 1909, it is said that officials recorded over 2,000 payments, while countless others went unrecorded. Hunters shot and trapped them, poisoned dens, and tore mothers from the forest floor with young still in their pouches.

People blamed thylacines for sheep kills, but later analyses revealed that feral dogs and poor husbandry were responsible for most losses. However, the paranoia was profitable. The bounty system made every carcass an income stream, every rumour an excuse for slaughter. As the numbers dwindled, curiosity grew. People captured a few survivors alive and exhibited them in travelling circuses or confined them to concrete zoo pens far from their natural habitat. Newspapers printed sensational stories, and captivated readers just as extinction closed in.

By 1936, protection could not alter the situation. That same year, the last known thylacine—called Benjamin—died of exposure in Hobart Zoo, pacing bare floors under floodlights. Grainy film captured his restless circling. The world's largest marsupial carnivore, tens of millions of years in the making, ended in a cage.

Its disappearance likely altered parts of Tasmania's predator–prey web. While research has suggested increased herbivore browsing pressure and competitive shifts affecting other carnivores, such as the Tasmanian devil, multiple factors—including later disease outbreaks—also shaped those dynamics.

The thylacine came to represent loss and possibility after its extinction, appearing on Australian emblems and postage stamps. Unverified sightings still incentivise expeditions into the Tasmanian wilderness, and modern de-extinction projects debate a potential reassembly of its genome. It died at the hands of bureaucracy as much as bullets and traps, and its absence still influences both the island's ecology and the concept of redemption.

Eldey's Last Stranglehold

Great auks once crowded the storm-beaten cliffs of the North Atlantic so densely that the rock looked as though it was feathered. A metre tall, heavy-bodied and unable to fly, they cut through icy currents, chasing fish with an aquatic grace that disappeared as they waddled ashore.

For centuries, their meat and oil fed crews and filled lamps; their down-padded bedding. When firewood ran short, the men burned their carcasses in the ship's ovens. By the eighteenth century, numbers dwindled, and scarcity turned the auk into a prize. Naturalists and private collectors vied for skins and eggs. The rarer the bird, the higher the price, and every rumour of a surviving pair drew hunters like a beacon.

In the early 1800s, the species retreated to isolated islands. Eldey, off the coast of Iceland, became its final sanctuary—yet even here the birds found no refuge. On 3 June 1844, three men rowed across the grey sea, following orders to acquire specimens for a merchant. They found the last pair guarding a single egg. With no defences against land predators, the birds offered no resistance. Strangled by the men, with the egg crushed underfoot, the last Great auks succumbed to the hunger to possess. Rarity only fed desire until the last became the most coveted.

Today, the Great auk's remains are inside glass cases. Skins and eggs sit in museum drawers, catalogued and labelled; their likeness immortalised in old watercolours and early lithographs.

Trophies and Trade

The extinction of these animals signals larger forces at work. As the preceding stories show, their endings trace the same currents that expanded our reach: trade, conquest and the hunger to possess.

High in Ethiopia's Bale Mountains, the Ethiopian wolf, Africa's rarest canid, still hunts rodents through frost-silvered grass. Only a few hundred survive. A single outbreak of rabies or distemper, or a campaign of poisoned bait, could wipe them out. It could have a similar ending to the thylacine's. An animal following its own path risks being drawn into ours and undone by it.

Passenger pigeons, the thylacine, the Great auk, and the river dolphin are signposts of a longer, tragic pattern. Each loss signifies the crest of a current that has shaped our species from the beginning of agriculture. Trade, conquest, and the urge to possess fueled cities, then empires. Looking back at these recent extinctions connects our oldest drives—curiosity, ambition, and hunger—with industrial instruments that have unmade the living world even as they built civilisation.

5

Ecology and Empire

Late Holocene–Early Modern (≈13th–19th centuries CE)

After European powers like the Spanish in Tenochtitlán and the British on the West African coast expanded their reach, the world changed rapidly as oceans became vital trade routes, facilitating the movement of people, goods, and even diseases such as smallpox, which played a major role in historical events like the conquest of Tenochtitlán. The Age of Exploration changed borders, belief systems, and the natural world.

Conquest sought control of peoples, of stories, and of nature itself. In its shadow, beneath rifles and royal decrees, extinction accelerated. Losses that had once been local and slow turned global and systemic. Animals that had evolved in seclusion for millions of years—on isolated coasts, plateaus, and forested islands—found themselves abruptly exposed to a predator unlike any other. On islands alone, nearly 30% of bird species are estimated to have vanished during this era. Many vanished before naturalists could even describe them; some before anyone realised they existed at all.

As fleets crossed the Atlantic and Pacific, those scattered disappearances fused for the first time in Earth's history. Every continent became bound into a single ecological circuit. No other creature had ever carried its crops, its livestock, its vermin, or its gods across oceans and planted them wholesale in alien soil. The European powers, driven by ambition and economic systems geared towards exploitation and expansion, alone stitched the planet together. In the same gesture, they unpicked it.

Maize and potatoes sailed eastward and fed millions, while wheat and sugarcane travelled west, carving fields from American forests. Coffee and bananas arrived in the New World; tobacco took root in the Old. Crops reshaped diets, while diseases unmade nations. Smallpox, measles, and influenza rode the same winds as the fleets and galleons.

Rats came ashore unnoticed, while missionaries deliberately spread religion. Goats, pigs, and cattle were intentionally brought to new lands by settlers seeking to transform and control the environment. This juxtaposition between intentional introductions and accidental stowaways underlines the varied ecological mechanisms at play. The Columbian Exchange was the widespread transfer of plants, animals, peoples, diseases, and technologies between the Americas and the rest of the world following Christopher Columbus's voyages in the late fifteenth century. There was a clash as ecosystems that had developed separately for millions of years mixed.

Some species flourished in the upheaval: wheat and cattle multiplied into new abundance. Planters transformed forests and grasslands into plantations. The Spaniards and Portuguese, the Dutch and English, the French and later the Americans—all carried the same template in their ships: tools, livestock, and expansion. Wherever they landed, they undid the web of life and rewove it in a modified form.

As Dead as a Dodo

On Mauritius, a green speck in the Indian Ocean, the dodo stood a metre tall: a flightless relative of the pigeon with a heavy, hooked beak and almost naïve confidence. When the first Dutch sailors stepped ashore in 1598, the bird had never encountered predators. It waddled among the newcomers with the indifference of a creature for whom danger was unknown. In a world without large predators, curiosity had never been costly—until the most dangerous one arrived.

Rats and pigs spilled from the ships' holds, raiding the dodos' nests for eggs. Sailors clubbed the ungainly birds for fresh meat or out of boredom. In less than a century, the dodo was gone; the last accepted sighting came in 1662. An animal that had survived untouched on its island Eden vanished before anyone could study it or even sketch it in life.

Long after its disappearance, the dodo gained a second life in the European imagination. Scraps of its remains—a head here, a foot there—made their way into museum storage. The dodo moved from existence to fable, from living presence to cautionary emblem. Sailors' tales of the odd Mauritian bird turned into legend back home. The very name "dodo" entered the lexicon as a symbol of obsolescence. The dodo's short, unguarded history on Mauritius prefigured the larger imperial story: abundance met by invasion, and then memorialised as a loss.

Where Fear Never Evolved

The tragedy initially unfolded on the islands, where it was most apparent. The clearing of over two-thirds of native vegetation on King Island, mainly for beef and dairy farming, dramatically altered its ecosystem and disrupted the original balance that had existed in relative isolation. Sailors found giant tortoises so docile and slow that they could be stacked alive in ship hulls to provide months of fresh meat. The very abundance of these islands was their undoing, consumed as provisions for fleets crossing the oceans.

On some islands, the first encounters ended almost as soon as they began. In the far north Pacific, a placid giant called Steller's sea cow was discovered by shipwrecked explorers in 1741 and hunted to oblivion by 1768. Each isolated paradise became a larder for conquering empires, and each paid the price for a harmonious existence.

What happened on islands in miniature was a rehearsal for what conquest would enact everywhere. At the far end of the world, when Europeans reached New Zealand, birds had taken the roles that mammals played. Owls, wrens and rails scurried in the undergrowth; giant flightless moas browsed the forest like deer. Soaring above them all was Haast's eagle—the largest eagle ever known, with wings spanning three metres. The settlers brought their own predators in their wake: cats and stoats, rats and new diseases. The widespread deforestation and targeted hunting of larger birds in Puerto Rico led to the extinction of several bird species within a few decades, including owls and flightless wrens. Earlier hunting had already caused the loss of the moas, and humans also caused the disappearance of Haast's eagle, a lineage that took millions of years to develop, in a few human generations.

Isolation, abundance, and naivete made species vulnerable when humans, seeking expansion, arrived. Equipped with advanced tools, beliefs, domesticated animals, and new ideas, imperialism consistently caused the same results: birds vanished from the skies, herds dwindled from the plains, and islands became silent.

Taxonomies of Loss

The losses of empire cut into the human imagination itself. Many parrot species that once lived in the Caribbean have disappeared, altering the relationships that communities like the Arawak had with these animals as ancestral guardians. Rituals collapsed; the forest itself grew less animate. In Australia, Aboriginal songlines mapped the land as a story. When Australian fauna dwindled, the songs lost part of their meaning.

Missionaries dismantled animistic beliefs, replacing them with catechisms. Settlers broke kinship systems and overlaid them with deeds and titles. Latin replaced indigenous names for plants and animals through the tongues of empire, erasing centuries of story. From the African savannahs to the plains of North America, whenever colonialism thinned the herds or felled the forests, the myths and meanings woven around those places unravelled as well.

The Age of Conquest offered the first glimmers of awareness while landscapes were stripped. Cabinets filled with skins, bones, and shells from every corner of the globe. Botanical gardens bloomed with exotic seedlings. Carl Linnaeus was a key figure in cataloguing the natural world, with his collections and classification work forming the basis of modern taxonomy. Collectors often prized rare specimens, especially as some birds and animals became increasingly scarce, making these trophies even more valuable. Knowledge and destruction advanced step by step, curiosity and violence bound. Cataloguing a species often hastened its disappearance, as demand for its skins and eggs rose as its numbers dwindled. We inherited maps and lists, taxonomies and theories of order—but we also inherited loss woven into the very fabric of understanding.

By the late 1700s, the concept of extinction had entered science: in 1796, the French anatomist Georges Cuvier astounded his peers by proving that fossil bones of mammoths belonged to creatures that no longer roamed the Earth. Once unthinkable, extinction was now undeniable—by 1812, Cuvier had documented forty-nine vanished species himself.

Once humans spread their system, it did not disappear. Further expansion made destruction faster. Technology continued this pattern, shaping the world we know today, though earlier empires also damaged the environment. But never before had the whole planet suffered at once. The age of sailing ships brought the world together, but that unity also led to global environmental degradation. Modern life was built as the natural world declined; we live with those consequences.

6

The Price of Bread and Circus

Late Holocene–Classical Antiquity (≈1st millennium BCE to 5th century CE)

It is easy to romanticise our ancient past: marble cities under Mediterranean suns, warriors beneath castle ramparts, explorers charting the unknown. I have been guilty of it myself, shaped by films such as *Gladiator, Troy*, and *300*. These narratives convinced me that this era marked humanity's ascent, but they obscure a crucial truth: our progress exacted a profound ecological cost.

The very foundations that elevated civilisation—technological and cultural achievements—came at the expense of nature, fueling its destruction.

This is history we can still touch: words etched in tablets, ruins weathered by time, and relics of a world already remade. Achievements like roads, aqueducts, governments, and the printing press are celebrated as triumphs, which in many ways, they were. Yet, the true cost—the erasure of the non-human world—remains nearly invisible beneath these legacies. While societies benefited, the environment paid the price. "Bread and circuses," the Roman phrase *panem et circenses*, meant little more than this: keep the people fed and entertained, and they will look away from the actions of those in power.

The Pleistocene had its hunters, small bands shifting the balance of established ecosystems. But with Rome, power was carried on its banners and spread through conquest, marking a new scale of environmental upheaval. Armies marched, roads cut through the landscape, and fortresses fixed their presence. Amphitheatres turned animals into entertainment, their deaths timed to applause. Fleets and fields required them to strip forests. Cities drew life from a continent, and people reworked rivers and coasts to feed them. Extinction here moved alongside legions, showing how military and cultural might produced unforeseen ecological consequences.

Rome's enduring influence can be seen in both its iconic aqueducts and amphitheatres, as well as in the preservation of its monuments by later rulers. According to Edward Gibbon, even the Gothic kings who succeeded the Romans were eager to protect these incredible structures. Perhaps that explains why many grown men, nearing thirty, succumb to a Roman obsession, as if it were a rite of passage.

Rome's hunger for animals filled the Colosseum with death. Bears from the Atlas, lions from North Africa, elephants from across the sea—brought for festivals that paraded exotic life to die for entertainment. These were state-sponsored displays. What vanished under Rome did so in chains, processions, and the roar of crowds.

The Last Bear of Africa

The Atlas bear, the only bear native to Africa, once lived in the Maghreb's mountains before being captured by the Romans for blood sports. Roman arenas often featured bears captured and kept in captivity to fight gladiators. The Romans confined wild animals in the Colosseum's underground chambers before releasing them onto the sand-covered arena as living targets. Venationes were public spectacles in ancient Rome that involved animal hunts and fights between men and beasts, held in amphitheatres alongside gladiatorial contests.

The Atlas bear became a prop in Rome's theatre, its death inflated the glory of emperors and fed the appetite of citizens who saw power made visible. It showed Rome's strength, proving that even the wildest creatures could be reduced to entertainment. The last Atlas bears survived only in scraps of memory. Folklore kept them alive in tales long after the forests were emptied, but their lineage had already broken. The first cages, taking off from the Maghreb for Rome, set their decline in motion.

The King Who Bled for Applause

The Barbary lion, with its heavy mane, was the prize of Rome's wildlife collections. Capturing a lion from North Africa meant capturing a symbol of kingship. Wild animals from across the empire were captured and transported to Rome, where large numbers of lions were displayed and killed in public spectacles to showcase imperial dominance over both nature and people. A lion tearing into a gladiator, or collapsing beneath a spear, was theatre that affirmed the empire's supremacy. Trajan's festivals were said to have slaughtered thousands of animals, lions among them, in a single sweep of imperial 'generosity' to satisfy the bloodlust of his citizens.

Roman hunts emptied habitats across the Maghreb, while the advance of farms and settlements cut into their range. By late antiquity, the lion had retreated from the Mediterranean world. While the Barbary lion was mostly wiped out by European hunters by the late 19th century, some evidence suggests that wild Barbary lions may have survived in Morocco until the 1960s.

Plumes for Power

The Arabian ostrich, which can reach two metres tall, once ranged across the deserts of the Near East. Its eggs were heavy, with a resilient shell, and its feathers were coveted as treasures.

The Romans made a fetish of the bird. Wealthy Romans often served ostrich eggs at their tables, used ostrich plumes for armour and regalia, and exploited ostriches for entertainment and display, including processions and arena spectacles. Archaeological evidence indicates that ostrich eggshells were used in ritual performances, which suggests that there would have been significant demand for these birds. Significant human population growth during the local Iron Age led to major ungulate extinctions in the southern Levant, which included the disappearance of the ostrich from the region.

Some survived into later centuries, lingering in Islamic art, in travellers' tales, in fading memory. The Arabian ostrich was an important part of Middle Eastern wildlife until it became extinct in the mid-20th century.

A Siege Engine Never Forgets

The Romans hunted elephants for their hides and ivory, but also found other uses for them by conscripting them for battle. Creatures with memory and intelligence were as much part of campaign plans as siege engines, their size and strength making them the ultimate beasts of war.

Long before the image of vast herds of savannah elephants entered the Western imagination, a smaller population of elephants roamed the forests and scrublands of North Africa. Known in history as the elephants of Carthage, these animals were taken from the region for centuries. Hannibal drove them across the Alps, a feat history hails as an audacious strategy. In hindsight, it was the utmost in cruelty. They forced animals, shaped by millions of years for heat and open plains, into icy mountain passageways, burdened them with armour, and had handlers prodding them. Most starved or froze on the mountain passes.

Following Carthage's fall, the Romans captured elephants and killed them in the Circus Maximus. Rome paraded them in triumph, reduced them to victory symbols, twisted their strength in mock battles, and forced them to trample criminals or gladiators for the audience. Rhinoceroses, rarer, suffered similar fates.

There were moments when the violence made even the bloodthirsty uncomfortable. In 55 BCE, Pompey staged a spectacle with twenty elephants. The crowd waited in anticipation of carnage. Instead, as the animals cried out, struggling against their fate, the audience faltered. Pliny the Elder observed that for an instant, the crowd pitied the elephants, troubled by the intelligence behind their cries. Though the sympathy vanished, it left a mark, and people realised what cruelty was, even in ancient times.

But the demand only grew. Roman ports overflowed with tusks and bones. Ivory and horn were carved into ornaments, furniture, and jewellery. Curiosity and conquest drove an appetite nature could not match; herds dwindled with every generation. We now know elephants mourn their dead, recognise themselves, and can even create art. The Romans used elephants only rarely and primarily for military purposes, showing little interest in the animals' intelligence or complex behaviours. They saw a military advantage cloaked in muscle, bulk, and durability.

Two thousand years later, the killing of elephants still continues. Elephants still fall victim to poaching for ivory, and rhinos for their horns. The weapons have changed, but the logic has not. Rome hurled spears; we point rifles. Today, an illegal ivory trade still thrives, and rhinos are butchered for a horn made of nothing more than keratin—the same material as a human fingernail—because some still prize it as a medicine. A startling measure of progress.

Rome vs. the River Horse

To the Romans, the hippopotamus was a thorn in the side of Roman expansion. Now considered the most dangerous animal in Africa, it trampled crops, sank boats, and loomed as an endearing relic of the conquered Nile. The Greeks had called it the "river horse," though it was closer kin to whales than to horses. Such subtleties had no place in the empire's view of nature. The wild existed to be exploited, displayed, or eliminated.

When the legions absorbed Egypt in 30 BCE, they inherited more than Cleopatra's treasuries and grain stores. They took possession of a living river. For millennia, the Nile had risen and fallen with the seasons, its floods feeding the soil and its creatures. Crocodiles and ibis guarded its margins; hippos churned its marshes. To the Egyptians, the hippo was divine and dangerous, an embodiment of the river's power.

Rome saw no divinity; it saw an entity of the natural world to be disciplined. The Nile's fertile silt was a resource for mouths across the Mediterranean. Wild floodplains became surveyed farmland. Canals, embankments, and irrigation schemes cut the delta into ordered grids as settlements expanded along the banks. A study on Roman-era Tunisia notes that animals introduced into new regions often faced threats from farmers attempting to protect their fields and from soldiers seeking sport, and from hunting parties eager to demonstrate their prowess. As human societies expanded in ancient Egypt, the habitat of hippopotamuses was increasingly restricted, and they were hunted, leading to their eventual extinction in the region. They were not driven to extinction as a species, but they had been driven from the river that the Egyptians had once built temples to their likeness.

The Romans saw only bulk, appetite, and nuisance. They never understood what they had displaced. Today, we know hippos shape the entire ecology of rivers. Their dung fertilises floodplains, feeding plants and crops alike. Channels are carved by their bodies, which keep waterways open. Fish populations are sustained by their grazing. When brought to Rome, it was displayed in the Colosseum during wild beast hunts.

Echoes from the Caspian

Rome never conquered Persia, but it pressed against her borders for centuries. The two empires—Rome in the West, Parthia and later Sassanid Persia in the East—locked horns in a long stalemate of raids, treaties, and ambitions. They traded silk and gold, arrows and hostages, rumours and animals. Their rivalry cast a long shadow, leading to the disappearance of an apex predator.

The Caspian tiger roamed the reed beds, river valleys, and forests from the Caucasus to Central Asia. It belonged to the borderlands of cold rivers and dry heat. It merged qualities of both India's jungle guardian and Siberia's snowbound powerhouse. Romans never saw it in the wild, but they saw it in cages.

Trade routes—some silken, others soaked in blood—funnelled exotic beasts westward. They paraded them through amphitheatres as emblems of conquest, then the crowds watched them being dispatched. To a Roman audience, the tiger was another adversary for a gladiator, another token of Rome's reach beyond civilian life.

Roman culture carried with it a language of occupation, absorbed by Persian elites already steeped in hunting as a mark of nobility. According to historical sources, the Roman demand for tigers and other exotic animals led to extensive hunting and trapping, which contributed to the decline of tiger populations in the regions where they were sourced. The most solitary of animals found itself without access to uninterrupted solitude.

The draining of the Mesopotamian Marshes was considered a tragic environmental catastrophe, and by the twentieth century, human activities such as railway construction and the introduction of firearms severely impacted the region's wildlife, which led to the eventual disappearance of species from the area by the 1970s. Today, the Caspian tiger survives only as preserved mounted skins and a limited number of incomplete skeletons. Its decline began during the Roman period.

The Price of Bread and Circuses

The amphitheatres are ruins now, their once full seats empty, their crowds long turned to bone and dust. The hunger to dominate, display, and consume the wild did not collapse with their empire. Today, we do not feed lions to gladiators, but we feed entire ecosystems to industry. The spectacle has shifted from the Colosseum floor to the slow violence of deforestation, bleaching reefs, and vanishing migrations, glimpsed in headlines and scrolled past without pause.

The Romans unveiled a way of seeing the world. To Roman eyes, this was man dominating nature, but it is important to see Rome on its own terms. These were not people armed with conservation science or ecosystem data, but with gods, myths and a need to survive. Most of its citizens were just trying to live their lives while being ruled under an iron fist. Their perceptions of the natural world were malleable; they were less educated than we are now, minds shaped by conquest and coin. In their view, the amphitheatre showed humanity's superiority to nature.

That mindset did not end with Rome; it continued and took on new forms. Steam engines replaced legions. Factories replaced amphitheatres. The exploitation remained the same; only the materials changed. The tragedy lies in the applause for the disappearance of these animals. Rome's spectacles ended in dust, but the worldview behind them survives, refined and globalised. The price of bread and circuses was not only blood. It was a silence that has grown louder with every age, and one that still surrounds us today.

7

The Weight of Innovation

Bronze to Iron Age (≈3000–500 BCE)

The Roman Empire is one of the best-lit stretches of antiquity. Its ruins still rise across the Mediterranean. Statues, inscriptions, and amphorae surface by the thousands. Coins carry the faces of emperors whose names can be traced year by year. Rome left behind a record dense enough to see how the empire stamped itself on landscapes and lives.

Step beyond that light, and the trail languishes. The written record reduced to silence. What remains is much harder to decipher: the clay tablets of Ugarit, the charred beams of Hattusa, the burial mounds and scattered pottery of prehistoric Europe. History becomes excavation, pieced together from remnants left by luck alone. In these ages, innovation reshaped societies primarily in physical terms, as humans learned to extract and master new metals, and these technological milestones set the stage for profound social transformation.

The overall story is apparent: as societies discovered and adopted new metallurgical techniques, such as those of the Iron and Bronze Ages, their structures and daily lives transformed. Archaeologists now read soils and bones, as they do with texts. From them emerges a picture of societies in transition. The next layer of our descent takes us into the Iron and Bronze Ages, when metals changed daily life, and survival depended on mining, smelting, and forging. These ages stretched for millennia. The span from the fall of Rome to the present is much shorter by comparison, making these ages crucial foundations for what followed.

Black Smoke and Red Metal

There is a moment in every life when power first announces itself; it can be taken and used. When I was nine, I realised I could convince my younger brother that his toys belonged to me. For humanity, that moment likely arrived with the forging of iron.

Reddish streaks in stone, black grains in river sands—iron was everywhere, yet reluctant to yield. To prise it from stone demanded extremely high temperatures. Ore had to be quarried, furnaces raised, and forests cut to charcoal. Bellows pumped without rest, flames coaxed to white heat, ore smelted into bloom. No single hand could master it. Entire communities bent to the task: miners digging, charcoal burners tending kilns, furnace keepers sweating in the fire's glow, hammermen shaping raw bloom into tools and blades.

The earliest objects—pins, knives, chisels—were tests of technique. But once the process took hold, iron remade the world. Blades kept their edge, and spears lengthened to reach further across the shield line. Armour grew stronger, shields more resilient. Weapons stock increased, intended for many rather than the few. The duel of champions gave way to the drilling of ranks. Triumph went to those who could marshal numbers, feed them, and keep them in line.

Iron appeared where ore, wood, and human willpower met. Assyrians in

Mesopotamia forged weapons for empires. The Nok in Africa smelted iron in furnaces independent of outside influence, reshaping landscapes and societies. In the north, iron ore was gathered from wetlands, hammered into blades and axes, fueling forest and fjord cultures. What makes the Iron Age strange is how it suddenly sprang up in scattered corners of the planet.

The difference in the manipulation of iron lay in scale. Unlike bronze, iron did not depend on caravans bringing tin from distant hills. Unlike gold, it was not a rarity used only for crowns or altars. Iron could be drawn from the soils of one's own land, smelted in village furnaces, and wielded by farmers as much as kings. The broad availability of iron made it the backbone of systems that demanded discipline and scale—miners at the pits, smiths at the hearth, soldiers drilled into formation. The animal that once hunted in small bands now stood in lines, shield against shield, its survival bound to the collective. This broad shift is central to understanding how metals drove social change.

The forge armed empires, opened soils, and drew forests into smoke. Upon mastering the ore, the path shut; returning proved impossible.

Appetite of the Forge

Iron's grip reached into many elements of life. It cut into the land itself. Iron ploughs tore deeper into the soil than wood or bronze could manage. Heavy earth that had resisted earlier tools yielded under iron. Forests fell, and fields widened. Harvests grew, and with them came populations. More mouths demanded more grain; more grain demanded more tools; more tools demanded more furnaces. Growth pressed outward in every direction, and the appetite of the forge matched the appetite of the field.

To free iron from ore required a heat that ordinary flames could not reach.

Charcoal was the answer, and charcoal devoured trees. A single furnace could swallow a grove. Across Europe and the Near East, pollen cores record the sudden retreat of oak and elm. The effects rippled outward. Wolves, aurochs, and wild cattle retreated as their ranges gave way to pastures. Once-common animals dwindled in middens, their bones replaced by sheep, pigs, and cattle bred for meat and labour. The human footprint crowded out rivals and reshaped allies. In Scandinavia, bog iron was gathered from wetlands, draining marshes and altering ecosystems that had sustained birds, fish, and amphibians for millennia. The illusion of progress masked ecological dependence. A forest could vanish in a season, a hillside stripped of trees washed into the rivers after heavy rain. The same tool that carved strength into human societies carved vulnerabilities into their foundations.

This was not a time of great extinctions, not yet. The aurochs still roamed, the wolf still prowled the edges, the forests still stood in patches wide enough to seem endless. Our post- Pleistocene impact was building momentum, and the patterns had begun to form. The Iron Age set the rhythm that would carry forward to the Roman Empire and beyond. Accelerating an unrelenting reduction of the wild under the pressure of human need. For the human animal, the change felt like mastery. For the rest of the living world, it was the start of retreat.

Selection for the Obedient

Iron altered human relationships, driving structural changes at every level of society. Ore had a more complex process for obtaining than the other resources we were used to gathering. Each step demanded bodies, skills, and discipline. A tool could be forged by one hand, but the system that produced iron demanded many hands bound to a common rhythm. This transition is the heart of the argument: new materials, particularly iron, restructured how people were organised and lived together.

Warfare exhibited the same characteristic. A sword could be wielded by a single fighter, but armies required soldiers drilled to move as one. Spears aligned in rows, shields locked in walls, formations that held under pressure. Battle was an exercise in logistics, supply, and command. Victory went to those who could synchronise thousands of individuals into a force of flesh and metal. Labour had to be timed, taxed, and counted. Rulers needed to know how much grain they needed to feed a garrison as well as how much they could take in the spoils of victory. The pressure of metal drew people into systems that demanded obedience, repetition, and routine.

This was a kind of evolution different from the biological kind that threads through most of this book. The selection of behaviours. Small bands of foragers had once rewarded cunning, boldness, and flexibility. Iron Age societies rewarded submission to structures. The freethinker could mutter at the edge of the hamlet, but people still ploughed the field. Even though the wanderer could slip away into the hills, armies still marched on. The traits most useful to survival were those that allowed an individual to endure command.

The timings of the forge, the drill of the army, the cycles of the field—these became the environments in which the human animal had to survive. Control exerted itself everywhere—from the commander, from the clergy, from the rhythmic pattern of effort and combat. Freedoms were consistently narrowing. On the other hand, these systems gave humanity new strength. They allowed large societies to cohere, to project power, to become robust. The cost was independence; the reward was hedging survival within the larger group.

Long before armies marched in lines and smiths tended bloomeries, humans had already bound themselves into systems. Making bronze required networks of exchange, record-keeping, trust and tally, the discipline of trade across distance. The ape who marched in formation had once followed the tally of rarer metals across mountains and seas. Iron hardened the systems that bronze had foreshadowed. Descending to bronze reveals our earliest relationship with alloy-making.

The First Global Alloy

Iron may have cemented systems of obedience, but bronze revealed their earliest outlines—the beginnings of large-scale organisation needed to work with rare materials. While easier to harvest than iron, bronze did not lie waiting in the ground. It had to be imagined into being. Copper gleamed in veins and riverbeds, soft enough to hammer into ornaments or blades. Tin was rarer still, scattered in distant hills, hidden in pockets far from fertile valleys. Combining the two conjured a substance that nature had never provided. This leap was both technical and social, since it required coordinated acquisition and distribution, foreshadowing later systems shaped by iron.

Two ordinary metals, when fused in a fire, became harder and sharper, enhancing their properties. The creation of bronze marked a shift in understanding. Materials could be changed, transformed, or combined into properties beyond their visible parts.

Copper and tin seldom occur together. To join them, people moved ore across deserts and along rivers, carrying it from one end of the known world to the other. Caravans crept through dust and heat, riverboats hauled their weight inland, and ships edged along the Aegean coast with the metal packed deep in their holds. Bronze belonged to no single valley or shore; every blade and vessel embodied the long routes that linked distant landscapes.

The cities of Mesopotamia reveal this web most clearly. Copper from Oman travelled north into Sumer, where it met tin brought from far-off lands. It was fused, then moulded into vessels, blades, and tools. At Uruk, scribes etched tablets of wet clay with tallies of grain, livestock, and labour. To track the movement of metal and food alike, they invented writing. Early documentation listed debts, responsible parties, and required items. Bronze sharpened both swords and abstraction.

Elsewhere, its forms carried power in different guises. In the Aegean, Mycenaean rulers were buried with bronze blades and ornaments, their wealth measured in metal. In China, the Shang dynasty poured bronze into moulds of exquisite detail: vessels adorned with dragons and masks, weapons etched with symbols of rule. These were offerings, but also assertions, metallurgy elevated into ritual. Across Europe, bronze axes and ornaments travelled along rivers and passes, their gleam moving into regions far from any mine.

Bronze was not as common as iron would later be. It did not arm populations wholesale. But it set the precedent. It showed that survival could depend on systems stretched across distance, on trust encoded in tally, and on metals that existed only when cooperation held. The ape that once lived by instinct now lived by exchange. Bronze was the first alloy of copper and tin, but also of landscapes, peoples, and the fragile webs that bound them together.

From Grain to Glyphs

Bronze gave our ancestors sturdier tools, but it also made possible the infrastructures that held early states together. Palaces and temples rose around storehouses where metal, grain, and labour could be counted.

Counting had been instinctive in animals long before. A crow can tell the difference between two and three. A chimpanzee can recognise a handful of bananas. But the leap into symbols was a significant differentiation. With clay tablets bearing marks, humans detached the number from the object. A wedge of clay could stand in for a sack of grain, a field of barley, or the body of a labourer. Reality could be summarised, then commanded.

Humans adapted to this new order as readily as they had once adapted to the savannah. Gods of fertility, storm, and war grew entangled with the language of tally and decree. Offerings were listed, sacrifices enumerated, rituals scheduled. Bronze also fixed hierarchies in place. Where the Stone Age had been more fluid, its bands shifting with season and need, Bronze Age societies had warriors and priests occupying the upper tiers. Farmers and labourers sank to their knees. The humans that once roamed now found themselves categorised, tallied, and defined by what they produced and what they owed. Bronze did not invent inequality, but it hardened it into a permanent condition.

For the inhabitants, this world resisted change, save for spiritual movements. Palaces endured for centuries, dynasties succeeded one another, and the alloy that underpinned it all gleamed without tarnish. The patterns of sowing, recording, taxing, and offering became part of life, inherited generation after generation.

The very systems that gave stability made societies dependent on them. If the tallies failed, if the shipments faltered, if the symbols no longer matched the fields or the stores, the illusion of permanence would dissolve. Bronze made the first successful civilisations possible, but it also revealed how tightly survival had become bound to abstraction.

Collapse of the Bronze Age World

The Bronze Age endured for between 1,500 and 2,000 years, its kingdoms trading across seas and deserts, its palaces filled with metal, grain, and tribute. Egypt, Mycenae, Hatti, Ugarit: powers linked by ships, caravans, and scribes. Envoys carried letters across courts, merchants moved goods along established routes, and scribes accounted for clay with debts and treaties. Then, around 1200 BCE, the weave came undone.

Cities that had stood for generations were burned. Mycenae fell, Ugarit vanished, and Hattusa was abandoned. In one kingdom, a desperate plea for aid was scratched onto a clay tablet and dispatched to an ally. The message survived while the city did not.

The cause remains heavily debated. Invasions by the so-called Sea Peoples, prolonged droughts, earthquakes, rebellions, the fraying of trade routes—perhaps all of them, converging in a storm of crisis. What is certain is the speed at which it happened. Within a few decades, palaces and temples that had dominated the eastern Mediterranean for centuries were left in ruin. Scripts disappeared, forgotten for centuries. The networks that had bound regions into a shared system collapsed, and with them the confidence that order was as permanent as their overlords had heralded.

We have clues of malfunction; charred beams from palaces toppled in fire, evidence of storage jars smashed and left in ash. Even tablet archives that end in mid-century, their records frozen at the moment when scribes stopped writing. In Greece, writing itself disappears for four hundred years, leaving only silence where once there had been tallies of livestock and prayer. The very interdependence that had made the Bronze Age powerful left it vulnerable. Copper from Cyprus, tin from Anatolia, grain from Egypt—cut one link, and the system strained. Cut enough of them, and the system snaps.

Pollen cores and lake mud preserve what the chronicles barely hint at. Hills once green with forest were stripped for fuel until the soil washed away. Drought rings cut through tree trunks; lakebeds record sudden shifts in vegetation that mark years of failed harvests and overworked fields. Trade routes faltered just as crops failed. The ground that had long masked its own strains gave in and withdrew its support. The networks that had seemed immovable could no longer bear their own weight.

For those alive in the twelfth century BCE, the change came as a jolt, rather than a slow fade. Buildings fell into abandonment, markets fell silent, and gods offered no reply. Writing itself disappeared from places where it had seemed permanent. Armies dissolved into scattered hamlets; survival shrank to the scale of soil and season. These people probably saw themselves as the height of civilisation—masters of bronze, of letters, of long-distance trade—yet when their networks cracked they reverted to older instincts, drawing once more on the hard cunning of their inner savage, a capacity for endurance that civilisation had merely suppressed.

Species vanish when their niches collapse; civilisations when the scaffolding of their order cracks. What failed was more than physical cities and kingdoms. It was the belief that symbols—treaties, ledgers, tablets—could anchor reality in place. Agreements lost their force, records turned to dust, and the abstractions that once bound the eastern Mediterranean ceased to function.

Human order is never a given, as it demands much, including steady landscapes, reliable exchanges, and trust that distant partners will keep their word. When strain overwhelms those conditions, the design unravels. The Bronze Age is an example of how interdependence can be pushed to the limit and then fail.

From the ruins emerged iron empires, stronger and more deeply rooted, humbled by failure. Even though we too depend on trade that spans oceans, on systems of exchange that feel permanent until the ground shifts beneath them. The same interdependence that once carried the promise of prosperity can tighten into fragility when climate, soil, or simple chance breaks the pattern. History shows that we are not promised a straight ascent: when networks fail, even the most advanced societies can find themselves stepping back toward the stone, forced once again to survive with the same elemental skills that pulled us from the wilderness.

Strip away metal, surplus, and memory, and what is left is the long apprenticeship that taught us how to be "human" in the first place.

8

Neither Beast nor God

Late Pleistocene–Holocene (≈3 million years ago to 500 BCE)

By the time we reach the Stone Age, we are no longer talking about kings or empires. Instead, we deal with traces of ancient humans that survive in DNA, soil, and cave art. This is prehistory in its rawest form—a world without written records or timekeepers. What remains are fragments, the faint evidence of a species becoming fully human.

The Iron and Bronze Ages, once vast, each spanned only a few thousand years; the Stone Age lasted millions. Almost all of our story happened then, before grain or metal. Archaeologists divide it into the Palaeolithic, Mesolithic, and Neolithic, but those labels are modern. For ancient people, life simply unfolded, unaware of distinct 'ages'.

Throughout most of the Stone Age, humans were not the only hominids on Earth. We were one group among many, moving with the seasons and animals, and just beginning to experience what we would consider complex thoughts. It was a time when we were neither fully animal nor fully advanced—still close to nature, yet already making symbols and tools that hinted at future human abilities.

Seeds of Permanence

The Neolithic began with seeds. A handful of grains was pushed into the soil—maybe by accident, or by imitating nature. Weeks later, a small miracle: green shoots rose where none had been before. That discovery was observed, repeated, and copied. In time, it became a bargain between people and plants. Wheat spread through the Fertile Crescent valleys, rice in Chinese wetlands, maize across the Americas, and taro and yams in New Guinea. Unique landscapes, different crops. Yet the pattern repeated, and people stopped moving.

We domesticated plants, but they also domesticated us. Wheat cannot spread itself, rice cannot irrigate, and maize cannot plant. These crops survived by depending on ours, making us their keepers.

Skeletons from early farmers show a decline: people grew shorter, spines bent, joints worn from labour. Teeth decayed from starchy diets; skulls bear marks of infection. People once lean and upright now bent over monotonous grainfields.

Foragers had lived as opportunists, moving when herds shifted, or rivers failed, their rhythms following the seasons and migrations. Farmers had no such luxury; neglecting the fields was not an option. Even abundance, when it came, carried its own dangers: more mouths to feed, more labour to rise for, more dependence on the next crop. This cycle only tightened with every generation.

Fields led to villages. Jericho built stone walls, among the first human defences. At Çatalhöyük, houses stood roof to roof, accessed by ladders; the dead rested beneath floors, and livestock were penned inside. This closeness let parasites thrive and made disease common.

Density bred division. A patch of land could now be claimed. A storehouse could be guarded and a surplus hoarded. What had been a fluid equality of mobile bands became a hierarchical society. Some families tightened their grip on grain while others invoked spirits or ancestors to sanctify their advantage. Property became an accumulation of possessions and something others could take from you.

Archaeologists find cracked skulls, arrowheads in ribs, and mass graves at settlements. People dug defensive ditches and raised timber palisades. Now, enemies were both predators and neighbouring settlements vying for fields during hunger.

Foragers lived by horizons; farmers by cycles: plough, plant, harvest, store. Surplus demanded counting: grain must be measured and watched. With storage came ownership, and with ownership, obligation.

Wealth inequality began influencing human societies over 10,000 years ago as social hierarchies formed. Families that controlled vital resources, such as granaries and fertile fields, gained significant advantages over others. The emergence of property also led to shifts in women's status within these changing economies, altering their roles in society. Fertility became a resource; bodies turned into lines of inheritance; even children, once the foragers' successors of the next generation, came to be seen as extra hands for the fields.

The wide-ranging mobility of the forager narrowed to a single plot of earth and the tasks it required. A failed harvest brought hunger, but it could also lead to debt or even the loss of freedom. The security promised by surplus also created new problems. Villages still rose and fell with environmental changes, and the abandonment of fields allowed forests to return. People could move, but not as freely as before. The shift to permanence was significant but not absolute. Humans had begun to shape their environments, and these new patterns of life formed a cage around them.

Between Freedom and Field

During the Mesolithic, glaciers retreated. Ice withdrew from the northern latitudes. Forests spread into the exposed ground. Rivers surged with meltwater. Animals returned to pastures that had been locked in ice for millennia.

For tens of thousands of years, humans moved constantly with herds and rivers, surviving scarcity by never standing still. As the climate warmed, the land stopped expelling us with winter, and game could be tracked nearby. The restless human began to pause.

At Star Carr in northern England, people built wooden platforms on a reed-filled lake. Deer skulls became headdresses, maybe for rituals now lost to time. Life was tethered—shaped by the lake, bound to its resources. Along the Danube at Lepenski Vir, stone houses rose, families returned yearly, the dead were buried beneath, and fish bones piled up. These were not yet villages, but significant, remembered places for ancient communities.

Wolves lingered at camp edges. Domesticated in the late Palaeolithic, their presence grew during this era. They became hunting partners and protectors, choosing to live with humans. As people gathered by water, bonds with canines strengthened. The Mesolithic remained a time of upheaval: sea levels rose, drowning lands like Doggerland, forcing people to higher ground, and old camps were lost to tides.

The Mesolithic mind, shaped by scarcity, began to consider abundance. Rivers and forests offered stability for new ideas—rituals, ornaments, a sense of permanence. Forager freedom remained, but with the weight of place.

Looking back, the Mesolithic feels like the last time humans were both animal and human—mobile, inventive, tied to wild ecologies, yet drawn to permanence. It was humanity's last pass of true animal-like freedom before agriculture.

Humans in the Long Ice Age

Picture forty people along a riverbank. Children chase each other, legs muddy, laughter and chatter echoing. Women carry infants, hands calloused from gathering. Men walk with spears, stone tips sharp. An old man limps behind, knees worn from hardship, but still keeping up. This was life for most of our species' existence.

The Stone Age spans about 3.3 million years; 95% of it is the Palaeolithic—an ocean of time versus the shorter Mesolithic and Neolithic. The Palaeolithic lasted through shifting continents, advancing glaciers, and the rise and fall of human cousins.

It is difficult for us, bound to short lifespans and calendars, to grasp such timescales. The rise of Rome to its fall fits inside a single heartbeat of that history; every king, every war, every book ever written occupies only its final pages. For millions of years, the human impact on the world was shaped by small bands following the herds.

Archaeology leaves us only the faintest outlines: rings of ash where hearths once burned, bones split for marrow, flakes of stone discarded in the dirt. From these traces, we glimpse the hints of existence: hunt, eat, rest, move. Firelight held the night at bay, and within its glow came stories—myths about animals, quarrels over food. Bands of this size did not require kings;

authority was fragile and easily dissolved by mockery or exile. Anyone who pushed too far was likely laughed out of the circle or killed.

This was what humans were accustomed to: a life of no lastingness, no limits. For them, home was a campfire, clothing was hide, and kin were everything. Palaces, borders, monuments, and gods were nothing more than a distant mirage of an incomprehensible future.

From Africa, bands of humans moved in a slow drift of tribes chasing food—following rivers and coasts. None of them thought of themselves as pioneers, yet within a hundred thousand years, humans had carried fire and flint to almost every corner of the globe.

Viewed from a distance, it seems absurd: a few thousand apes with spears somehow traversing the entire globe. Motivated by their hunger, multiplied across time and place, they secured a future that eventually led to the reshaping of the planet's biosphere.

This dispersal marked the origins of humans becoming a planetary force, through the steady accumulation of ordinary acts—fires lit, carcasses hunted, rivers crossed. A single footprint in the mud was the beginning of a global reach. By 10,000BCE, only a handful of islands remained untouched, and the humans that had once clung to a few remote valleys in Africa now stretched across the entire Earth.

Life in the Palaeolithic was brutal. A failed hunt brought hunger that deepened with the snow, and a broken leg could turn a person into a burden that winter or other tribe members would not forgive. Fire was a thread that bound life to the living, and a single flame might decide whether a band survived the night or froze.

The wilderness never gave our ancestors room to rest; a sickness could sweep through a cramped shelter, a quarrel could destroy the trust that held a band together, a lean season might scatter them into the snow. What archaeology leaves us are the faint remains of those failures—stone tools abandoned by floods, footprints halted in volcanic ash, skeletons half-buried where camps once stood. Our appetites, our vigilance, our reflex for story all carry the mark of those millennia of cold. History likes to begin with grain, with cities, with kings. But those are new inventions built upon minds already tempered by ice, minds long practised at survival in landscapes vast enough to erase us.

The Stone Age and the Pleistocene describe different things, and it is easy to blur them together.

The Stone Age is a human story. It refers to the period when stone was the primary material for tools, from about 2.4 million years ago until the first widespread use of metals. The divisions archaeologists provide, the Palaeolithic, Mesolithic and Neolithic, are important for discussing cultural phases.

The Pleistocene is a geological term. It marks the most recent Ice Age, beginning about 2.6 million years ago and ending about 10,000 years ago, defined by the repeated advances and retreats of the great continental ice sheets.

The two periods overlap almost perfectly: for most of the Pleistocene, our ancestors lived in the Stone Age. That is why the terms are often used as if they mean the same thing. Yet one describes the planet's climate and landscapes, the other the technologies and lifeways of the people who lived through it.

As we move forward, we remain within that Pleistocene world—still in the Ice Age itself—but we now turn from human survival to the wider environment they shared: the shifting ice, the changing forests, and the other species that thrived and disappeared in that cold.

9

The Planet with Frostbite

Late Pleistocene (≈70,000–11,700 years ago)

Measured against present-day comforts, and even those of recorded history, the Pleistocene was an entirely unfamiliar way of existence. Bronze and iron brought granaries and shelter; even the lean seasons of Rome unfolded behind walls and beneath a calendar. Step further beyond those few thousand years, and the civilisational safety net falls away. For most of our species' time on Earth, there were no markets, no storage, no written law—only small human bands navigating a planet that forgave nothing.

Glaciers advanced from the poles in slow catastrophes of weight, bowing the crust and carving valleys to the bedrock. Seas retreated more than a hundred metres, baring continental shelves now under the ocean. Winds lifted flour-fine dust and spread it across continents. Ingenuity helped, but resilience was essential to survival.

At the peak of the last glaciation, around twenty thousand years ago, ice sheets two kilometres thick covered much of what is now Canada and northern Europe. A wall of white bent the light and made weather of its own. South of it stretched treeless plains, their grasses stiff with frost. Herds of mammoth, reindeer and bison set the pace of survival: lose their trail and the season was lost.

Generations were born and buried without seeing the ice retreat. Later civilisations counted time in dynasties; here, time was measured by the slow creep of glaciation, each beat drawn out over thousands of years.

The Clockwork of Ice

Earth's path around the Sun stretches and tightens over spans of 100,000 years, and the planet's tilt shifts by a few degrees on a 41,000-year rhythm. The axis itself traces a slow wobble over about 20,000 years, so that the timing of the seasons drifts. These Milankovitch cycles alter the distribution of sunlight, and, though each change is imperceptible over a human lifetime, together they determine whether the north melts or freezes. When summer light weakens, winter snow lingers, and the planet's bright surface reflects more sunlight back into space.

The atmosphere magnifies the effect. When the oceans draw down carbon dioxide, the air holds less heat and the ice advances. When carbon dioxide rises again, warmth gathers, and the glaciers retreat. The planet oscillates in these long movements: a 100,000-year inhale of cold followed by a slow exhalation of warmth. Ocean sediments and polar ice cores record over twenty such swings during the Pleistocene. To those alive then, each lifetime passed beneath an unchanging horizon of frost. Greenland's ice layers do show sudden warmings of several degrees within a decade, triggered when floods of meltwater freshened the North Atlantic and slowed its great conveyor of heat. These planetary rhythms set the stage for everything that followed at ground level.

Today, the Holocene, our current warm interval, began about 11,000 years ago, when the great ice sheets retreated from Greenland and Antarctica.

Their withdrawal left the soils that now feed the world's grain belts and the meltwater rivers that guide our cities. Major advances such as farming, villages, and states emerged only after the ice ages ended and exposed new land. The stability we experience is but a brief chapter in Earth's much longer geological history.

The Planet at Low Tide

The outlines of the continents during the last glaciation's peak were nothing like those we know now. With so much water locked in ice, sea levels dropped by more than 100 metres. Britain and Ireland formed the western hills of a broad lowland, Doggerland, which connected them to Europe. Siberia and Alaska were joined by the dry grasslands of Beringia, a land bridge wide enough for herds and small human parties to reach the Americas.

South of the ice sheets stretched a belt of grass and herb that palaeoecologists call the mammoth steppe—a pasture on the scale of continents. Cold, dry air kept forests at bay, while short, bright summers drove explosive bursts of growth. The animals and the predators that fed on this system turned the steppe into a living engine, its story to be taken up in the next chapter.

Elsewhere, forests retreated into pockets: oak and beech clung to Mediterranean peninsulas, spruce to refuges in the American south. The planet's wind belts shifted, and monsoons weakened. Even far from the ice, the atmosphere and oceans bent to its gravity. Glacial time seemed slow, but it could move suddenly. When the last Ice Age ended, the Younger Dryas plunged the north back to near-glacial cold for more than a millennium before warmth finally returned.

The Gamble of a Thousand Ancestors

Against this planetary machinery, human life was stripped to its bare essentials. Bands of perhaps twenty to fifty people moved continually. Their entire economy could be carried on their backs: flint blades, bone awls, cordage, embers banked in ash. A broken spear could mean hunger; a lost ember, death by cold. Shelter was whatever the land allowed—caves, rock overhangs, or frames of wood or mammoth bone covered with hides. Clothing was technology as vital as any weapon: layered furs sewn with bone needles. Food came from a shifting mix of roots, nuts and small game; large kills were rare wins, not certainties. A failed season meant famine. There was no surplus in the agricultural sense—no granaries to buffer a lean year.

The cold never let up; hypothermia could kill in hours on any day. Frostbite turned our most dexterous tools into liabilities. Before antibiotics, a flint cut or tooth abscess could become a death sentence. Predators stalked the same trails—cave lions in Europe, giant bears in the north, wolves everywhere. Humans were part of a much larger food web.

Survival depended less on individual toughness than on collective endurance. An individual could not absorb a broken ankle or a fever; a band could. Anthropologists note that such groups rarely exceeded about 150 people—the limit of memory and trust. Within the circle, children were raised by many hands, elders remembered the routes of migrating herds, and meat from the hunt belonged to everyone. The grandmothers who helped nurse children were just as vital as the strongest hunter.

Some genetic models suggest that the population of our *Homo sapiens* ancestors dropped to around 1,280 breeding individuals for approximately 117,000 years. The human story nearly ended at one of the narrowest points in our lineage.

At the time, they could not recognise that they were our species' important few. They were coastal foragers, waking to the same urgency each day: find food, guard the young, outlast the night. The dangers they faced were ordinary—yet the stakes were absolute. A single season of misfortune could have eradicated us from the world.

Every civilisation, every language, every idea that followed depended on their unyielding persistence. The art of Lascaux, the mathematics of Babylon, the music of Bach—all of it lay unimagined inside that tiny population. They kept the entire future possible.

Billions of living humans are the consequence of a survival so narrow it remains almost invisible in the archaeological record. For hundreds of thousands of years, we were mobile, communal, precarious. The environment forged bodies and minds adapted to uncertainty. We learned to plan far ahead, to track absence and presence, to imagine herds not yet seen and seasons not yet turned. Cold demanded foresight; scarcity demanded cooperation.

Those reflexes remained even when barley flourished and walls arose. We carried them into villages and, later, into cities; you carry them now: the constant checking of the horizon, the uneasy feeling of what may be beyond the dark while you fall asleep.

10

Life on the Mammoth Steppe

Late Pleistocene (≈100,000–11,700 years ago)

Imagine a horizon that refuses to end. No treeline steadies the eye; no ridge interrupts the sky. Underfoot lies rocky ground, pulverised by glaciers and carried on winds that howl for weeks. In spring, those winds strip the surface bare; in summer, they settle again, laying down a mineral wealth that waits beneath the frost. Snow falls but vanishes by midsummer, feeding roots that have withstood months of darkness. The hills presided over a tight weave of sedges, grasses and flowering forbs. Each spring, this tapestry bursts into a green tide that rolls from west to east.

This was the Mammoth Steppe: a single biome stitched like a broad belt around the top of the world. From the Pyrenees to Alaska and across the Beringia land bridge, it linked Eurasia and the Americas into a single connected pasture. By some estimates, its productivity rivalled that of the modern Serengeti, despite the cold. Ice ground mountains into silt, silt fed grasses, and grasses bound the soil and fed herbivores. The herds themselves closed the loop—hooves churned the earth, dung fertilised the roots, and seasonal migrations carried nutrients across continents. Remove the grazers, and the grassland falters; remove the grass and the soil drifts back to dust.

The scale of this frozen wilderness would have been disorienting. Herds moved in their thousands, dark bodies sweeping across pale grass. Predators shadowed the edges of migration. Their own paths were written into the same soil as their prey. This was a grand ecosystem, driven by the tension between cold air and fertile ground, and by the animals that meshed the two.

Shapers of the Mammoth Steppe

At the heart of this ecosystem stood the woolly mammoth, taller than any living elephant and wrapped in a double coat of underfur and long guard hairs that protected against the Pleistocene chill. Their tusks cut through the crusted snow to reach the shoots beneath. A single herd could reshape acres of steppe in its passage, trampling paths into ice, toppling shrubs, and scattering dung that seeded the soil. By keeping the land open, they prevented the forest from creeping back.

Beside them moved steppe bison, bigger than their modern kin and armed with metre-long horns. Where mammoths opened the land, bison scoured it. They cropped grasses close, stamped nutrients into the ground and pivoted between Siberia and Alaska in their yearly migrations. The aurochs—ancestral wild cattle—travelled in smaller groups but had a similar impact on the territory they roamed.

Broad-shouldered and low-backed, the woolly rhino preferred solitude. Its coat was the colour of rust; it moved alone through the frozen valleys. Its long, flattened horn swept aside the snow to reach the grasses beneath. They lived alongside herds of Irish elk (*Megaloceros giganteus*), whose antlers could stretch over three metres across. Antlers of this size were proof of the steppe's abundance: minerals in the soil, calories in the grass, summers long enough for bone to grow into forests of antlers. Despite the name, these deer were not true elk, and their range extended well beyond Ireland. These deer were creatures of the steppe, with a range stretching from Ireland to Siberia.

Every herd lived under the gaze of predators. Dire wolves (*Canis dirus*) tracked the bison, their jaws deeper and stronger than those of any living wolf. Like modern wolves, they likely hunted in coordination; fossils of entire packs fill tar pits. Other predators joined them. Long-limbed and almost ten per cent larger than modern lions, cave lions (*Panthera spelaea*) were stripped of savannah manes. They raised their cubs in stone hollows and hunted across the open grass. Ice Age artists painted them on cave walls, and some of their bones lie close to human hearths, evidence of competition for shelter.

Twilight of the Giants

The steppe stretched across half the planet's northern face. Dust from the steppe crossed oceans. Herd migrations redistributed nutrients on a continental scale. But scale did not guarantee permanence. Orbital cycles tipped back toward warmth, so grasses yielded to shrubs, then to trees. The herds thinned. Predators did the same. By the time the last woolly mammoths lingered on Wrangel Island, the great pasture had already dissolved.

The close of the Pleistocene was a saw-toothed retreat—brief warmings, sudden returns to cold. The fauna of the steppe had weathered such shocks before. Mammoths and bison had survived the undulating warmth and centuries of deep freeze long before the first spearhead glinted. They were built for flux, with wide ranges and robust adaptations.

As the last great glaciers ebbed, their abundance faltered. The world that had seemed unassailable came undone because an additional force entered the pattern. When our own species stepped into the north, the calculus of survival changed. Hunters crossed Beringia and fanned out through Europe and the Americas.

Here lies one of the most argued questions of the Ice Age: what ended the reign of the megafauna? Was it the warming planet, the human hunter, or a convergence of both? The evidence carries both signatures. The timing is unnervingly close. The Mammoth Steppe had endured every climatic whip-crack the Pleistocene could deliver so far. Nevertheless, not long after our arrival, the great herds dwindled.

11

The Paleo-Economics of Collapse

Late Pleistocene–Early Holocene (≈12,900–11,700 years ago)

About 12,000 years ago, the long ice age that had endured for over 2 million years came to an end. Across the planet, the largest mammals—creatures that had outlasted eight major glacial cycles—vanished within a few centuries. Only Africa, long home to hominids, kept its bigger animals. The mammoth steppe had dissolved into a world of smaller ecosystems.

The Pleistocene climate was challenging, shifting between extreme cold and sudden periods of warmth, but these difficult conditions alone did not eliminate megafauna from the plains; studies have found that declines in megaherbivore populations coincided with warm, wet periods. Species now extinct had survived worse. They endured abrupt warming of several degrees, glaciers advancing, and droughts that turned lakes into saltwater. The end of the Ice Age brought no especially harsh climate. Yet the giants still fell in unusual numbers.

Palaeontologists call this the Late Quaternary extinction.

The case rests on scattered sites where extinction timelines overlap with human arrival and direct physical traces of contact remain in the ground.

The Campo Laborde site in South America is the only confirmed evidence of humans hunting giant ground sloths. While in Australia, diprotodonts and giant kangaroos disappeared within a few thousand years of the earliest human fires. Some evidence suggests the devastation of coordinated culling.

Imagine the planet as a vault, accumulating biological riches over inconceivable spans of time. In the briefest moment—a geological instant—those riches were spent, emptied by the arrival of a new kind of hunger. This is the paleo-economics of collapse: the planet's slow-grown life, millions of years in the making, squandered in a geological heartbeat.

Was it the abrupt slip of climate—the Younger Dryas cold snap and the warming that followed? Was it hunting—the sudden spread of a tool-making primate whose ingenuity multiplied the force of its muscles? Or was it the convergence of both: a world already in flux met by a predator that learned too quickly? No single answer survives scrutiny without the others. The evidence speaks in layers, more so than declarations: genomes that show long declines; kill sites littered with fluted spear points; ecological models proving that even modest hunting of adults could drive populations below recovery levels once habitats are fragmented.

The giants did not fall due to a sudden catastrophe from the sky, nor were they felled by climate change alone. It is difficult to have this conversation without mentioning the new predator who carried forward every lesson learned and improved upon it until the accumulated weight of culture became a power greater than any other living creature.

The New Kind of Predator

Wolves had long hunted in packs; lions had long stalked in the grasses. But a different take on the hunt surfaced among them. The first spark coaxed from stone, the understanding of seasons, the angle of a spear's fluted groove—none of these vanished with the one who first discovered them. They passed from voice to voice, each retelling sharpening the lesson. Knowledge compounded like interest. Over generations, understanding became a second inheritance layered upon genes, a force that evolved faster than biology and shaped the environment as surely as ice and wind.

Across the landmasses, the pattern repeated. In Eurasia, hunters carried the atlatl, a carved shaft that extended the human arm and hurled a spear with lethal speed. Clovis flintknappers in North America cut a channel at the base of their points so a spear could penetrate more deeply and be held more securely. In Australia, Aboriginal peoples practised fire-stick farming: burning corridors to encourage plant growth, to herd game, and to reset the landscape in their favour.

The steppe's lions and wolves hunted much as they had for ages; their strategies were bound by instinct. Our own lineage broke that limit; each kill added to a growing archive of tactics and tools. Culture turned the act of hunting into an accelerating technology, and with it, an eighty-kilogram primate became a destabilising presence in ecosystems that had persisted through millions of years of turbulence.

Vulnerable Titans

Across the broad plains of Eurasia, the mammoth herds had known every cold the Pleistocene could deliver, but never the presence of humans. When the first hunters crossed the steppe, carrying tools the giants did not yet recognise as danger, they had no answer for survival. In the shallow basin, a band of hunters waited, bodies lean from winter, eyes fixed. The mammoths smelt the air, and then surged; calves were forced into the centre, adults on the circumference. The air was filled with the bellows of terrified animals.

For the hunters, the danger was immediate; a single error and tonnes of muscle could crush ribs to splinters. Archaeology captures these encounters in evidence from across the world. At Lange–Ferguson in South Dakota, mammoths were felled and butchered where they dropped. At Murray Springs in Arizona, fluted Clovis points still lie among the bones. Across Eurasia, entire shelters were built from mammoth ribs and skulls—evidence that the harvest of giants was a pattern.

A single mammoth held over two million calories—enough fat and protein to sustain thirty people through the darkest winter month. Marrow was drawn from broken limbs, fat rendered for light and heat, hide turned to shelter and sinew to cord. No smaller prey could match such a return on investment. Hunting rabbits or deer for the same yield would have demanded hundreds of kills and days of pursuit. The arithmetic of survival favoured targeting the larger animals.

Mammoths likely had gestation periods of 21 to 22 months, similar to modern elephants. This slow rate of reproduction meant that hunting could have had a devastating impact by outpacing the species' ability to replenish its population. The strategy is clear in the evidence. Hunters learned to drive herds into bogs or box canyons, to set fires that funnelled animals into kill zones. In some places, cliffs bear the blackened soil and butchered bones of mass kills. Selective targeting of prime adults—the reproductive core of a species—likely occurred.

Each victory fed a band of humans. Every success also lowered the herd's survival threshold. What looked like an inexhaustible treasury was already running at a deficit. The megafauna had outlasted droughts, abrupt warmings, and the rise and fall of ice sheets. When they met a new predator—an ape who carried memory through successive generations and multiplied power through shared understanding—the animals perished. To the mammoth, the spears were only the surface of the threat. Behind them stood knowledge and calculations, which turned risk into certainty and abundance into collapse.

The Climate's Last Whiplash

Around 12,900 years ago, just as the great ice sheets were in full withdrawal, the Northern Hemisphere snapped back to near-glacial conditions. This thousand-year relapse—known as the Younger Dryas—was sudden and sharp enough to test every species that had tracked the thaw.

Most researchers trace this reversal to the Atlantic itself. As the Laurentide Ice Sheet melted, torrents of fresh water poured into the North Atlantic, disturbing the great ocean conveyor that carries warm equatorial currents north. Shut down that flow, and the northern hemisphere cools. Ice cores from Greenland record temperature plunges of several degrees within decades—fast enough to be felt within a single lifetime.

The typical Pleistocene pattern had long included abrupt swings, but the Younger Dryas stands out for its timing. It arrived just as ecosystems were re-sorting themselves following thousands of years of ice. Habitats that had opened for grazing closed again; forests retreated, grasslands spread, then forests expanded once more as the cold relented.

Not everyone agrees that ocean currents tell the entire story. In the early 2000s, a group of geologists proposed a more dramatic trigger: fragments of a comet striking Earth, sparking continent-wide fires and filling the air with soot. Their evidence includes thin layers of dark sediment rich in soot and rare elements such as platinum, along with microscopic diamonds and magnetic spherules that can form in high-energy impacts. To them, these signs point to fire from the sky and a sudden planetary chill.

Some experts contend that each of these markers can arise without a comet.

Wildfires leave soot, volcanoes scatter magnetic spherules, and ordinary geological processes can redistribute the unusual metals. Replication has been uneven; some sites show the signal, others none at all. The comet hypothesis is fascinating but unproven, according to most Quaternary scientists. The prevailing explanation remains the simpler one: a surge of meltwater and the abrupt re-shaping of ocean circulation.

Whether driven by currents or cosmic debris, the Younger Dryas was the last great cold snap of the Ice Age. As the Younger Dryas waned and the Holocene warmed for good, this period coincided with the global presence of our own species.

Taken together, the evidence points to a double pressure rather than a single fatal blow. The climate at the end of the Pleistocene lurched through sharp reversals—abrupt cold snaps and sudden warmings. Meanwhile, our own species had spread across the continents, hunting with tools and tactics that multiplied their impact. Whether a comet ever fell, ecosystems still strained under natural flux, and human predation tipped them past recovery. The convergence of climate change, shrinking refuges, and the extinction of the giants ended the mammoth steppe and much of the Ice Age's megafauna.

12

Primordium Mentis

Pliocene–Pleistocene (≈4 million–40,000 years ago)

The origins of the story of our impact, seen through the Ice Age, begin earlier on a seam in the Earth: the Great Rift Valley of East Africa. A scar that cuts through Ethiopia, Kenya, and Tanzania, it is here that *Primordium Mentis*—the origin of the mind—takes root. In this fissure, shifting plates, swinging climates, and unstable landscapes combined to produce something unprecedented: an ape that could think beyond reflex and instinct.

The Rift was tempestuous during our ascent. Rains filled vast lakes that drew animals to their shores, only for the water to vanish in a generation, leaving life parched in the aftermath. Forests spread, then gave way to grassland, which in turn yielded again to thickets and scrub. Adaptations that worked in one season could fail in the next. In such a place, brute strength was not enough, and speed meant little. Survival belonged to those who could change their behaviour as quickly as the surrounding land changed its face.

In this environment, memory and imagination became as important as teeth or muscles. Behavioural plasticity—the ability to learn, improvise, and pass those adjustments on—was the evolutionary prize. A single mind could solve a puzzle, but a group of minds, sharing strategies, pooling discoveries, and remembering together, could bend the odds in their favour. It was the ability to connect intelligences that set us apart.

The conditions of the Rift are still with us, carried in the wiring of our brains. Our appetites were shaped by scarcity, and scarcity is written into us. Sweetness once meant we had found an energy source to survive another day. Today, when calories overflow, the same craving remains blind to its new environment. The same goes for the urge to know who was loyal and who might betray you. It was the difference between being defended at a carcass or left behind when the leopard struck from the canopy.

Finding meaning in chaos once meant avoiding the randomness that could kill. Imagining predators where there were none was safer than to dismiss a shadow that concealed teeth. These instincts are ghosts of the Rift still threading their way through our desires, our politics, and our everyday lives. Civilisation has buried them under layers of culture and invention. We patch on systems of law, technology, and ritual, but the foundation remains mostly unchanged. We still carry a mind tuned to that older world, a cognitive engine built in scarcity and sharpened in fear. Our brains may have installed the latest modern life update, but make no mistake, we are still running it on a Pleistocene operating system.

The Other Humans

The forests and valleys of Ice Age Europe held more than dire wolves and cave lions. They held another kind of human. A genetic cousin, we call them Neanderthals. They called themselves nothing at all.

They weren't the half-gorilla, half-man monsters conjured in Victorian lecture halls. They lived in small bands, close-knit nomads. Flint flakes still lie buried where they worked. Their brains matched ours in size, and sometimes surpassed them. Like *Homo sapiens*, they were built to utilise vision, memory, and muscle for survival.

But Neanderthals did not begin in Europe. Like us, their story began in Africa.

Homo sapiens is believed to have evolved in Africa, with the oldest known remains found at Jebel Irhoud, Morocco, dating to about 315,000 years ago. This evidence suggests that the species was widely distributed across Africa by then. Another moved north into Eurasia, adapting to new climates and strange winters. Time stretched, changes accumulated, and that path became Neanderthal.

Their frames were stockier, their faces more pronounced. They grew attuned to survival in the cold. Where we travelled quickly, they rooted deeply. For over 300,000 years, they shaped lives that conformed to their terrain. Their territories stretched from the Atlantic to Central Asia. Their world was thought to be a patchwork of smaller bands, often isolated, sometimes joined. Knowledge moved more slowly, carried from one hearth to the next.

The question that shadows them is why we survived while they faded. Populations shrank unevenly, withdrawing towards the edges of Europe. The last bands held on in Iberian caves. Climate stress, small population size, and encounters with us all played a part. Perhaps our networking, desert and coast adaptation, or an elusive factor secured our advantage. The answer may be less a single cause than the accumulated weight of differences.

Neanderthals left no songs, no words or stories that survived the collapse of their kind. But they left marks through genetics. Genes are mute; they cannot recall the cadence of voices or the weight of silence between them. They were another kind of human, running in parallel with us for hundreds of thousands of years. Their end was the closing of an entire possibility.

It is tempting to imagine what history would have been like if it had tilted in a different direction. Had the Neanderthals pulled through while we slipped into extinction, what world would stand in place of ours? A civilisation with no Homer—poet or Simpson—no Einstein, no Christ. Instead, names unknown to us, carried in languages that will never be spoken again. Myths rooted in valleys we no longer know of, laws declared in ways that would make no sense to us.

They didn't entirely disappear, but they did not survive as a distinct species. The genetic record confirms this beyond speculation: fragments of Neanderthal DNA live on in us. Had a plague cut through the first bands of sapiens, or a refuge favoured Neanderthals during climate shifts, the roles might have reversed. Modern humanity could have been Neanderthal in body and mind, carrying only a faint echo of sapiens in its genes.

They were the last of a genus that once held many kinds of humans. Now we move across the Earth as the only branch still living, carrying fragments of their blood. It is easy to mistake our presence for inevitability. The Neanderthals endured for 300,000 years, a little less than we have survived so far.

What potential world vanished when theirs fell silent? We cannot know, but under the weight of our rise, the question of what could have been remains.

The Ghosts in Our Blood

In the depths of Denisova Cave in Siberia, a child's finger bone lay hidden beneath the soil. For tens of thousands of years, it rested there in the dark, as glaciers advanced and retreated, as mammoths and cave bears came and went. When it was uncovered in 2008, no one guessed it belonged to a branch of humanity that had never been named. From a fragment no bigger than a coin came the revealing of an entire people lost to time: the Denisovans.

Denisovans are known more through blood than through bone. Millions alive today carry their ancestry, though most are unaware of it. Their DNA lingers in Tibet, where a gene passed down allows lungs to draw oxygen more efficiently in the thin air of the plateau. Where Neanderthals can be traced in pigments and shelters, Denisovans persist stitched into our cells.

With the discovery of the Harbin skull, the Denisovans stepped out of the shadows, though only just. We can picture their stance, their strength, and their endurance across harsh landscapes, from the Siberian steppe to the high Himalayas. Denisovan genes appear across Asia, through Oceania, and as far as the Pacific islands. They dispersed, interbreeding with Neanderthals and with us. Their genomes prevail in people from Australia, Papua New Guinea, and Southeast Asia. Unlike Neanderthals, whose traces cluster in Europe and western Asia, Denisovans left a broader inheritance. Their extinction is harder to trace. Populations may have dwindled slowly, absorbed by others, erased by time. The limited number of bones makes it difficult to reconstruct the full story.

This makes them different from Neanderthals, who left pigments, tools, and hearths. Denisovans confront us with an absence. Beneath our skin, Denisovans remain. And in that persistence lies a warning: what they were to us, we may one day be to others.

The Fire Starters

Homo erectus means “upright man,” but “enduring man” might describe them better. For nearly two million years, they walked the Earth—a span ten times longer than *Homo sapiens* has lasted so far. Their tenure dwarfs our own.

They carried the echo of an older world in their faces. Heavy brows jutted over deep-set eyes, jaws were both broad and powerful, the skull long and low with no chin to soften the line. They would have looked more archaic than us, closer to the primates from which we came. Nevertheless, the body beneath was relatively modern in form. Long legs, narrow hips, and skilled hands gave them an ease of movement and a capacity for craft. Their faces bore the past, while their stride pointed to the future.

In the Rift Valley, the land baked under the sun, grasses rippled to the horizon, and rivers carved their channels. From there, *Homo erectus* spread outward, through forests, across deserts, along mountain ridges, and onto islands cut off by seas. They reached as far as Java, long before we would invent the sail. Erectus was the first human with a global range, not tethered to one habitat but fitted to many.

In East Africa, in caves in Israel, and at open sites in China, layers of ash and charred bone show flames ignited with intention almost a million years ago. Fire reshaped existence; it guarded against the night, drove back predators, and softened food so that more energy could be drawn from it. Cooked diets lightened the load on jaws and stomachs, freeing energy for brains. But the gift of fire went beyond function. It stretched the day into the dark, drew bands into circles of light, and created hours where thought and memory could compound.

The Acheulean hand axe was a sharp, symmetrical, balanced, and durable tool. For more than a million years, these teardrops of stone accompanied erectus across continents. Culture here was preserved with continuity—memory in the hand, knowledge carried forward through muscle and imitation.

The structure of their brains suggests the capacity for speech, though we cannot know for sure. Perhaps they used grunts and rhythms, gestures and calls, a proto-language sufficient for the *Homo erectus* world. Meaning may have moved among them, seemingly enough to hold a band together.

At Dmanisi in Georgia, one skull tells a story of community. It belonged to an elder who had lost every tooth. The jawbone had healed, meaning life continued long after chewing became impossible. Survival in such a state demanded long-term assistance. Food must have been chewed or softened, shelter provided, and protection given. In that healed bone lies evidence of life extended because others shared the burden. Even in a harsh world, space was made for the vulnerable.

Erectus was built for covering enormous distances. Their long legs and lean frames were adapted for the chase. They hunted by driving animals to exhaustion, wearing them down over hours. They spread across continents by the same method, stepping across seasons and generations until the Old World was filled with their presence.

Through the millennia, their populations rose, dispersed, and gave rise to others. In some places, they became *Homo heidelbergensis*; in others, Denisovans; and likely in other species yet to be discovered. The branch that remained in Africa would, in time, become us. Their fire still burns in our hands, their hardships live in our stride. Even so, their two million years still came to an end.

Descent From the Trees

Between four and two million years ago, in the shifting landscapes of Eastern and Southern Africa, a genus of small-bodied creatures bore an unlikely biomechanical signature. We call them Australopithecus. Their bodies were dainty, weighing little more than a modern eight-year-old child. Their skulls were small, with brains the size of a chimpanzee's, around 400–500 cubic centimetres. By most measures, they belonged to the forests from which they had come, but their posture told another story.

The pelvis was short and broad, tilted to carry their weight upright. The spine found its distinctive S-shape. Knees angled inward, so each step brought the feet beneath the body. The big toe aligned with the others, no longer grasping for branches but for planting into the earth. These were early experiments in bipedalism. They could still climb trees—their arms long, their fingers curved, their shoulders strong—but the ground was no longer forbidden territory.

It was a dangerous gamble, as two legs were slower than four. They were less stable, more prone to tipping, and harder to twist or turn in escape. Upright posture made them visible on the savannah, their silhouettes stark against the grass. Standing upright did not make them any stronger, but every bipedal step was a hands-free step. They could carry food, gather branches, and hold infants close. They could scan further across the horizon, watching for danger. It wasn't all advantages; childbirth grew more challenging, and narrow hips strained to deliver large-headed infants. Back pain, hip strain, and joint injuries became the price of walking. The cost was high, yet the negatives failed to surpass the positives.

Climate shifted violently across their millennia. Forests gave way to savannah, and rivers shrank and swelled. Herds of zebras and antelope moved with the seasons. To survive here, adaptability was essential, and Australopithecus embodied it. They could climb for safety, forage in the trees, but they could also range across the plains, spreading further than their ancestors had ever gone.

They were not a single species but many, appearing across time and place: anamensis, afarensis, africanus, sediba. Each blended the old with the new, some more primitive, others edging closer to what would later be called Homo. Living in groups, their lives were brief and vulnerable. They ate fruits, roots, seeds, and leaves. They scavenged when they could, always at risk of losing what they found to the claws and teeth of more powerful animals.

The shift to walking was pivotal in helping our ancestors survive on the savannah. Without it, hands could never have been freed to strike stone; fire would never have been kindled, nor children born with larger brains. For Australopithecus, upright posture altered how the body moved through the landscape, yet it remained bound to fragile ecologies, limited diets, and narrow climatic tolerances. The success of their line eventually disintegrated into extinction, their forms gone from the earth.

Not a single species survived to see what their gamble allowed. But in the fossil record, one of them would speak louder than the rest. Her bones were buried in the soil of Ethiopia, resting for 3.2 million years until the ground gave her back to us.

Under the Diamond-Filled Skies

In 1974, in the Afar desert of Ethiopia, a team of paleoanthropologists brushed back the earth until the outline of a skeleton emerged. Ribs, a pelvis, a jaw, and long bones from arms and legs.

That night, music drifted across the team's camp—Lucy in the Sky with Diamonds. The name Lucy stuck to the discovery, nameless in life but assigned one, millions of years after her death. She had walked, foraged, and fled through her brief life, no different from the rest of her band. The circumstances of her death ironically granted her immortality.

She was tiny, just over a metre in height. Her hips and knees showed marks of a walker. The strength of a climber was evident in her arms and shoulders. Her body lived in two domains at once, balanced between tree and ground, past and future.

Her own end was likely ordinary; the bones suggest a fall. Nothing in her story marked her for remembrance. She remained because conditions around a riverbank preserved her long enough for her eventual discovery.

Lucy's skeleton has since become an icon; her remains have travelled across the entire world. Her bones sit in cases now, measured, scanned, and reconstructed. But before she was a fossil, she was alive. Fear was in her chest, and hunger in her belly. She knew the warmth of others near her, the stress of a predator's movement in the grass. She breathed, as you are breathing now. The world she crossed was different, yet the rhythm of her life was not foreign. You know it already in your own body.

Under the same diamond-filled skies that would eventually give a song its name, Lucy walked. She carried no sense of consequence, no glimpse of the vast chain of futures suspended on her shoulders. Her kind was already moving along a collision course with extinction. She did not know, and could not have known, that we would one day look at her bones and call them our own beginnings.

Beyond the Dawn of Humanity

Our species now stands alone, but the ground beneath our feet is filled with the remains of countless others who once lived as we do. What we call our history is the faintest residue of an inheritance almost entirely erased.

Approximately twenty-five hominin species have been identified. There may have been a hundred more. Most are gone without name, without record. Their faces will never be reconstructed, their footsteps will never be tracked.

We could dedicate chapters to the hominids that almost were, in every branch of our tree—jawbones lined in cases, skulls reconstructed under glass, footprints burned into ash. Entire libraries of books already do this.

But the human story alone cannot tell the larger story of what led to the rise of our species.

For all the weight we have given to our origins, they are late pages in a far longer book. The world turned for billions of years before us. Oceans seethed, mountains split and joined, forests rose only to fall, and extinction erased what life had written.

The human chapter is brief and ends here.

The ground holds older memories.

And the bones beneath us go far, far deeper.

13

Continents Adrift

Paleogene–Quaternary of the Cenozoic (≈66 million years ago to present)

In the last chapter, we touched on the Pleistocene, where mammoths grazed, and cave lions prowled alongside our own ancestors. Those were glimpses closer to our time, fragments of a world still tethered to the human story. Now the measure of time changes, and from here, we will regularly cover spans of millions of years.

We arrive within the Cenozoic, the last sixty-six million years, and the age of mammals. It began with the Palaeogene, about 66 to 23 million years ago, and continued through the Neogene, from 23 to about 2.5 million years ago, before leading into the Pleistocene Ice Ages and on to the present. Within these broad epochs lie the Oligocene, Miocene, and Pliocene, each marked by long transformations in climate and life. The next three chapters move backwards through this expanse, tracing the gradual reshaping of continents and the emergence of ecosystems and animals that give the Earth its modern character.

Here, continents themselves become the protagonists. Landmasses drift across the planet, colliding or breaking apart. Collision brings strangers face to face, reshuffling and rewriting the order of life.

This is the scale on which the next chapters unfold. We will not follow the precise ticking of geological clocks; the span is too vast. Instead, we will pause in past worlds made strange by time.

South America in Isolation

Three million years ago, the landmass that would come to make up South America stood apart. The continent housed kingdoms, shaped by isolation that lasted tens of millions of years. Oceans had sealed it away since the age of dinosaurs, making it an evolutionary island on a continental scale.

This long solitude gave rise to animals that seemed improbable. But every one of them was real. Isolation allowed them to follow strange trajectories, freed from the pressures that shaped life elsewhere.

On the broad plains that would one day be Argentina, herds of glyptodonts grazed. They were the size of small cars, protected by a reinforced shell, with short, heavy tails. They moved low to the ground, their armour shielding them from attack. Among them lumbered *Doedicurus*, a similar animal with even greater bulk. At the end of its tail grew a mace of solid bone, rimmed with spikes. When swung, it could crush a predator's ribs or break a rival male's shell. These were mammals, but their forms looked much more ancient, closer in spirit to the armoured reptiles that had ruled in a bygone age.

In the rivers, *Toxodon* waded through reeds. Its body carried the weight of a hippo, its teeth ground like those of a rodent, and its snout was broad and blunt. Darwin himself, when confronted with its fossils, admitted confusion. He could not decide whether it was related to pigs, hippos, elephants, or rodents. Today, ancient protein and mitochondrial studies place *Toxodon* as a close relative of horses, tapirs, and rhinos.

Farther across the plains, *Macrauchenia* roamed in groups. They stood on slender legs, bodies tall and upright, necks long, their faces ending in a short, flexible trunk. Their silhouette echoed a giraffe, though smaller. They lived in a world without large carnivorous mammals, but not without predators. South America had its own rulers.

Here, apex power belonged to birds. The terror birds, *Phorusrhacids*, strutted across these lands. Some were smaller, no taller than a man, but others towered over two and a half metres. Their heavy legs were built for speed. Their beaks curved into hooks sharp enough to split flesh. Their wings had dwindled to stubs—useless for flight but powerful for balance as they sprinted across the grass. They filled the role that lions and tigers would take elsewhere: the ultimate authority of the plains.

At this point, early hominids were still confined to Africa. Australopithecus walked the savannahs, *Homo habilis* struck stone for the first time, but their lives were distant from each other. South America remained untouched by their presence until later. The continent disregarded our ancestors, existing independently.

This was stable for millions of years, but isolation never holds forever. Beyond the horizon, tectonic plates were closing the ocean gap. A bridge of land was rising grain by grain, rock by rock. Eventually, the barrier that protected South America would falter. Predators and prey from the north would march across, each carrying its own evolutionary strategy. Wolves, cats, bears, and horses were coming.

The Isthmus Land Bridge and the Joining of Two Worlds

For over 30 million years, the Americas' fauna evolved in relative isolation. North America carried herds of horses and camels, predators shaped in dialogue with Eurasia, and packs of canids that spread across its plains. South America followed another path, sealed away behind vast seas. Each continent nurtured its own order of life.

The Isthmus of Panama is the narrow land bridge that now links North and South America. For tens of millions of years, currents surged through the seaway that split the continents. Then, beneath those waves, sediments accumulated. Volcanoes rose, islands flared and collapsed, and the sea-floor heaved upward. It took millions of years, and no animal alive at the time would have been aware of its progress. Generations passed, extinctions swept the board, new species rose and fell, while the waters grew shallower and the gap grew narrower.

Once the two continents were connected, exchange began almost immediately. From the north, predators pressed southward—wolves, sabre-toothed cats, bears, foxes. Behind them came waves of herbivores: deer spreading through the forests, horses galloping across open ground, pigs rooting in new soil. From the south, giants began their own march: ground sloths as heavy as elephants, and the armadillos that had swollen to monstrous sizes.

Northern carnivores cut deep into the ranks of the southern fauna. By the time the northern big cats arrived, most of the largest terror birds were already in decline; multiple factors drove the turnover. Even the crocodilians that had held sway along the rivers receded into backwaters. All the same, the southern migrants still made their presence felt in the north. Ground sloths lumbered into North America and prospered for a time, reshaping woodlands with their bulk. *Glyptodons* advanced as far north as the grasslands of what is now Florida. For several millennia, they mingled, their lineages interwoven across a new, shared landscape.

The record usually reduces this moment to charts. Arrows crawl across the Isthmus. Names sort into columns: arrivals, survivors, extinctions. But on the ground, it was less abstract. It was a wolf following the unfamiliar scent of a sloth, or it was a terror bird locking eyes with a cat it had never seen before.

Ambush in the Pampas

Two million years after the Isthmus closed, the Pampas lay beneath a red horizon. Light pooled across tall grass that jostled in the wind. By a rocky outcrop that broke the unending grasslands, *Megatherium*—the largest sloth to ever exist grazed, with their massive claws curled inward as they dug up roots, tore bark from trees, and stripped shrubs bare. Dirt and moss clung to their shaggy coats. Each adult weighed over three tonnes, so they moved slowly, with the certainty of creatures that, once fully grown, had never feared becoming prey.

But new predators had wandered into this ecosystem.

In the grass beyond, two *Smilodon* moved inwards, low and coiled like sprung traps. Their bodies were sinewy through the shoulders, armed with forelimbs built for grappling mega-herbivores. Their skulls gaped wider than any modern cat, jaws expanding to drive sabres down in a single, efficient killing strike. They were predators honed in another world, carrying weapons the sloths had never evolved to defend against.

One sloth paused, the mindless rhythm of feeding broken. Its nostrils flared, pulling the air in short bursts, ears twitching. For a heartbeat, the group remained calm until the grass split open.

A flash of amber burst forward. The first cat hit the side of the sloth, claws raking into flesh, anchoring itself against a wall of moving muscle and fur. Dust rose in a cloud as the giant and the predator collided, the sound carrying across the valley. The second *Smilodon* broke from cover, circling for an access point, each step calculated, eyes locked on the throat.

The sloth bellowed, a sound deep and guttural, rarely heard on those plains. It reared upright, claws swinging through the air, in a last attempt to rid itself of the cat. But the *Smilodon* clung on, dragging its weight into the lungs of the giant, forcing it towards exhaustion.

The second closes in.

The *Megatherium* staggered, knees folding. The sabres struck, driving into the neck. Bone snapped, throat pierced. The final bellows collapsed into a whimper. Vultures circled above, summoned to witness the inevitable. The promise of a scavenger's jackpot would soon be theirs.

The sloths had grown enormously in a world sealed away from predators of this kind. The cats had crossed the bridge from the north, their hunting strategies honed from battles with other giants.

The Island That Drifted Alone

Around 45 million years ago, Australia broke from Antarctica and began its northward run. Over the next ~10 million years, the Tasmanian Gateway deepened, severing the last shallow links and helping lock in Southern Ocean circulation. While South America was headed for a later collision with North America, Australia drifted alone.

For tens of millions of years, isolation shaped its fauna. Ungulates, lions, wolves, elephants—none ever established here. Marsupials diversified into roles filled elsewhere by placental mammals; monotremes endured in their ancient egg-laying niche. The continent wasn't sealed, though: bats were present by the early Eocene (~54–55 Ma), and murid rodents arrived from Asia in the late Miocene (~8–5 Ma), long before people, and later dingoes changed the story.

Australia's soils are among the oldest and most exhausted on Earth. Rain erodes quickly from its rivers, leaving basins dry and stubborn. Seasons here never followed reliable patterns; drought and fire dictated survival. Vegetation grew tough, roots deep, leaves sharp—poised for the harsh climate. The animals that lived on it had to adapt to scarcity, to diets of fibre and spikes, and to landscapes that could burn to ash in a week, taking decades to return. In such a place, evolution favoured resilience, patience, and peculiarity.

This is why the continent raised creatures unlike any elsewhere. *Diprotodon*, the largest marsupial to ever live, lumbered across the plains as a grazing giant whose bulk shaped the landscape in the same way elephants shape the savannahs of Africa. *Procoptodon*, the short-faced kangaroo, stood two metres tall. Unlike the whimsical springing of modern kangaroos, it likely advanced with a heavy, two-footed stride.

Predators came too, though not in the forms one might expect. *Thylacoleo*, the marsupial lion, weighed little more than a leopard but carried a bite as strong as any mammalian carnivore relative to its size. Unlike the cats of Eurasia, it reared its young in a pouch. Catlike, but not quite anything we would recognise today. Then there was *Megalania*, the giant monitor lizard, six metres from snout to tail, dragging its scaled body through scrubland, flicking a forked tongue to scent the air.

For all its strangeness, Cenozoic Australia was a coherent system. Its ecosystems functioned, tightly bound to the soil and climate that shaped them. Isolation, though, carries fragility. When the barrier breaks, when outside forces intrude, the system cracks. Australia's strength came from its uniqueness, but so did its weakness. Species honed in solitude were inadequately armed against newcomers. They had evolved without placental carnivores, without wolves or big cats, without the fast-breeding herbivores that spread elsewhere. In time, as additional pressures arose from human arrival, this world would also unravel.

Australia, then, is the counterpoint to South America in the story of drifting continents. South America tells us what happens when once-isolated faunas collide. Australia shows what happens when they never collide at all. It creates entire ecologies that seem, at first, outlandish yet are logical within their own frame. It also serves as a warning: the more unique a world is, the more vulnerable it becomes. To see what isolation created, we need to move once again from generalisations and principles into the lives themselves.

Tension in the Pleistocene Dust

Three million years ago, in what is now the Northern Territory of Australia, a *Diprotodon* carcass lay half-submerged at the edge of a waterhole. Its flanks split open, while the flies gathered, teeming around its mouth and eyes. The outback heat is already accelerating the putrefaction, the air carrying the iron tang of decay.

The water stirred; a ripple widened, faint at first, then insistent. Out from beneath the undercut bank came a head, broad and angular, mud-flecked across its scales. *Megalania*. The lidless eyes were fixed on the carcass. Shoulders broke the surface tension of the water, then the torso, followed by the tail dragging behind. It paused, flicked its tongue into the air, and moved closer.

It tested the *Diprotodon* with a single claw, widening the wound. The skin yielded immediately, and the scent intensified. Its jaws gaped and saliva frothed, rows of teeth catching the light.

From the bushes, another presence.

Thylacoleo, the marsupial lion. It stepped out of the heat haze, its body low, its gait stiff. Each claw flexed, half retracted. Its eyes cut across the distance to *Megalania*. It stopped twenty paces from the kill.

The tongue of the lizard flicked again. The marsupial's breath is slow and controlled. The flies continued to swirl over the *Diprotodon*, their buzz the only sound in this moment.

Megalania surged, prepared to battle for this prize. *Thylacoleo* lowered its head, determining if the price was worth paying. It leaned forward. One more step, and the fight for sustenance would unfold.

It seemed inevitable.

Then the marsupial lion blinked, breaking eye contact. It pulled its claws back into their sheaths. The calculation favoured surrender. The carcass was not worth the risk. *Thylacoleo* returned to the lizard's gaze a second longer, then slipped back into the scrub, vanishing into the bushes as if it had never been.

Megalania remained and tore into the flank. The sound of flesh parting mixed with the violent thrashing of the water.

On this continent, rivalries were resolved through the constant calculation of risk. Where energy was scarce, bodies were specialised, and gambles could cause more problems than they set out to solve.

Isolation and Collision

South America and Australia reveal two distinct outcomes of continental drift. One shaped its destiny through collision, the other through isolation. Both were written by a planet that moves on scales far beyond human perception.

Continents are only lines on a map to a handful of human generations. In reality, they are conglomerations of plates of rock, drifting on currents of molten mantle, colliding and parting in rhythms set by heat deep within the Earth. The pace is imperceptible; in the span of a human life, they seem permanent. In timescales of a million years, they behave more like a liquid, reshaping everything in their path.

This is the rhythm of deep time. It outpaces the rise and fall of species, outlasts even the boldest lineages. Australia's solitude and South America's collision are part of a cycle in which oceans open, mountains founder, and continents move all over the globe.

These intervals belong to a scale where the unit is the constant turnover of entire ecosystems. What feels everlasting within human memory is reduced here to a passing adjustment, a brief arrangement in a story that knows no urgency.

14

Blueprints of the Familiar

Miocene–Pliocene of the Neogene (≈23–2.6 million years ago)

By the Neogene—roughly twenty-three to 2.6 million years ago—the continents had moved into positions close to those we see today. Seaways narrowed and basins shallowed. The late Pliocene uplift of the Isthmus of Panama severed the last broad exchange between the Pacific and the Atlantic, reshaping global ocean circulation. Currents reorganised; heat transport changed course; the climate took a cooler path. Far to the east, the collision of India with Eurasia pushed up the Himalaya and the Tibetan Plateau, while the slow press of Africa into Eurasia added height to the Alps and Atlas. High ground reworked wind, rain and rivers. The atmosphere began to feel drier, with colder seasons and sharper winters, and with starker boundaries between wet and arid lands.

The mid-Miocene held a warm interval. Then the long slide into cold continued. Ice gathered on Antarctica and advanced toward lower latitudes. Where winds drove along coasts, cold water welled up, lifting nutrients toward the light. Plankton responded in great seasonal surges. On land, forests opened. Light reached the ground and stayed. Through these couplings, a new global pattern took form. This age worked through persistence, through accumulation. The outcome: a planet that cooled and dried across many regions. With that, cooling and drying brought a world supporting a different landscape logic—one increasingly ruled by grass.

The Empire of Blades

Grass only looks ordinary because we live inside its victory. Measured against the green dynasty of the grasses, Genghis Khan's empire stands as a brief, local skirmish. During the Neogene, it expanded from regional provinces into a planetary order. A lineage within the grasses evolved the C_4 photosynthetic pathway, which excels under intense sunlight, low carbon dioxide levels and periodic drought.

Between about eight and two million years ago, C_4 grasses spread across the warm, seasonally dry belts of Africa, Asia and the Americas. On the interior plains of North America, they came to cover open ground by the late Miocene and early Pliocene.

Comparable shifts unfolded across Africa and parts of Eurasia. Where forests had once held sway, savannahs, pampas, steppes and prairies took root. The plants carried their own tools of conquest. Dry stems fed the fires. Flames swept across the open ground, clearing saplings and cycling nutrients. Soon after, grasses sprouted from growth points near the soil. The burn-regrow rhythm became part of the grassland's standard operations. Over time, these systems helped cool the planet. They drew down carbon effectively in hot, open settings. Burns brightened the ground. Dense roots bound the soil, stored carbon and governed how rain entered and left the system.

Animals answered the new script in step, with entire lineages of browsers lowering their heads to the ground and becoming grazers. Teeth rose into high-crowned molars able to endure the abrasive grind of silica-laden forage. Skulls stretched to house larger brains. Stomachs gained specialised chambers where vast microbiomes fermented cellulose and liberated energy that no vertebrate enzyme could accomplish alone.

The hidden relationship between herbivores and microbes became as cohesive as muscle and bone. Limbs lengthened and lightened; tendons took on more of the work, acting as springs that store and return energy. Herds travelled along paths cut by water and wind. Hooves broke soil; seed-heads hitched rides in fur; fruits moved through guts and fell farther afield in dung. In the wake of these movements, grass tillered into thicker mats, bare patches opened for opportunists and trampled corridors soldered one habitat to the next.

On open ground, concealment becomes harder. The margin for error narrows. Canids extended their stride, refining cooperative pursuit across long distances. Felids refined the use of fast, explosive bursts of power. Many shifted their ambush to thickets, gullies and river margins, where cover still granted a few seconds. Sabre-toothed cats developed upper canines shaped for deep, targeted cuts. Hyenas joined as hunters in their own right. They became consummate bone-crushers, able to extract calories from what remained.

Proboscideans—the order that includes elephants, mastodons and mammoths—reshaped woodlands with their bulk. Alongside them moved other giants, like the giant ground sloths of the Americas. *Paraceratherium*, a distant relative of the rhinoceros and the largest land mammal ever known, left its mark on the landscape. A passing herd deepened wallows, widened waterholes and pressed leaf-litter into soil. Where proboscideans roamed, shade retreated and light broadened. Horses shifted between three toes and one; camelids roamed North America long before their later dispersals; and giraffids tried both extreme necks and heavy headgear. Deer are divided between forest and grassland life, while bovids diversified into grazers and browsers that shared the same ranges in terms of height and season. When the Pliocene cooled and the climate grew erratic, many of these specialised lines disappeared. North American horses narrowed to a single dominant branch, eccentric proboscideans vanished and some camelid families faded from their birthplace.

Over these millions of years, mammals reshaped to match the new world of grass. Those changes were the first outlines of the climate we still inhabit. The slow interplay of grass, fire and migrating herds drew carbon from the air, cooled oceans and began the long oscillation of ice and thaw that frames our own history. What we call the modern world—its seasons, its patterns of rain and drought—was built here, in the unending work of species and landscapes that eventually provided stability.

Oceans in Parallel

The sea answered the same planetary changes with its own order. Upwelling strengthened along coasts where winds drove surface waters away from shore, drawing cold, nutrient-rich water into the light. Narrower gateways and altered wind fields shaped new gyres, concentrating productivity into bands that drifted with the season. Earlier cetaceans hunted fish and squid with teeth, while baleen whales developed a new way of feeding. A single lunge through a dense patch of krill delivered a payload measured in tonnes.

Before the late Neogene, most baleen whales were modest in size by modern standards. As productivity settled into reliable flows, selection favoured ever-larger frames capable of moving between seasonal hotspots and storing energy on a larger scale. Blue whales and their kin are the visible summit of that trend. Sperm whales dived into deep water after squid, storing oxygen in muscle and blood and using sound to map the dark. Nearer the surface, a macro-predatory sperm whale—*Livyatan*—carried large conical teeth in a skull built for ramming. It likely hunted sizeable prey, including other whales. In the same seas, a shark perfected to exploit a world rich in marine life cruised. *Megalodon* reached lengths of a bus and bore triangular, serrated teeth. The mega-predators' fortunes rose with the productivity of the sea and faltered as climates cooled and prey distributions shifted.

Marine abundance rested on delicate rhythms. When currents weakened or winds fell still, plankton blooms waned, and the food chains they fed began to disperse. Whales adapted with journeys that spanned the planet, banking fat in one ocean and spending it in another.

In the late Miocene, the continental shelves lay drowned beneath warm, slow-moving seas. Today, those drowned plains are beds of silt and limestone, where divers bring up a scatter of teeth, each as long as a human hand. Such concentrations mark a nursery: a place where the largest shark that ever lived began life away from the worst of its own kind. Modern great whites use similar havens—bays shallow enough to keep out larger hunters, rich enough in prey to feed pups already two metres long.

Even giants start small and need a safe upbringing. A gravid female slid in from the deep to one of these shallow areas, her body heavily pregnant, widening through the mid-section. She slowed, tasting the water. The chemical trace of a turtle drifted in—as sure a signal as any beacon.

Nearby, a Miocene sea turtle lifted for air. Its shell was as broad as a dining table. For the shark, such prey offered dense, compact calories. Fossil shells from these coasts still carry bite marks from predators of her kind. Her calm movement creased the surface. The jaws opened, three rows of teeth at the helm. She took a bite and shook her head; the water filled with chalky grit and the spray of red. In a few slices, the shell collapsed into her mouth. A clean kill and an easy meal. She drifted broadside for a moment, belly full, tail moving her through the nursery, other younger *megalodon* and mothers nearby. Mackerel sharks bear live young that grow in the womb by consuming unfertilised eggs; the pups are born large but still vulnerable and these shallow, food-rich flats stack the odds.

Blueprints and Beginnings

By the close of the Neogene, the continents stood in the outlines we know, and the oceans had settled into the circulations that still feed their fertility. Under those slow pressures, the planet took on the familiar shape. Against those spans, the few centuries of our own upheaval feel abrupt. Earth's systems bend and recover, but only on clocks that run far longer than any human history. What we call stability is the long, unhurried sum of countless adjustments.

In these pages, we have followed mammals as they came into their own—testing forms, establishing the foundation for the lineage that would one day face its own perils. But the story of mammals does not begin here. To see how those foundations were first laid, we must understand the volatile events that gave mammals their chance. The Neogene offers the comfort of a world we can recognise; the next chapter returns to the dawn of the mammals.

15

The Inheritors of Hell

Paleogene—Palaeocene–Eocene (≈66–23 million years ago)

The Palaeogene stretched from 66 to 23 million years ago, an epoch when mammals shaped the Earth in their image. Their origins, however, lay deep in the age of the dinosaurs. For more than one hundred million years, these creatures lived small and hidden, with bodies no heavier than a loaf of bread. Warm blood kept them alive through the chill of the night. Quick hearts drove them to forage under cover, slip through leaf litter, and survive by remaining unnoticed. Above them, dinosaurs ruled the ecosystems. Mammals felt these titanic forces through tremors in the soil and the sudden stillness of a forest after something much larger passed.

Then the sky delivered a single event that defined the end of the dinosaurs: an asteroid from deep within the cosmos collided with the Earth, ending their era and opening every path that would one day lead to us.

The Last Perfect Day

The first sign of destruction was a dot of light in the Cretaceous night sky, as faint as a star against the dusk. Evenings passed. The speck grew, brightening and swelling. It shed a tail that trailed like molten hair across the sky. Animals may have lifted their heads at the glow, puzzled, then returned to grazing. Day by day, it expanded as it neared the planet's atmosphere.

By the final dawn, it dominated the sky–a stone the size of a city, wrapped in its own glow of fire. It came on in silence at first, but as it drew near, noise arrived as shock more than sound. A crackling shriek detonated into a roar.

When it struck the Yucatán Peninsula off what is now Mexico, the sea turned to steam in an instant. The Earth's crust burst open and flowed like molten metal. Shockwaves raced ahead of sound, flattening forests hundreds of miles away. Incandescent debris burst upward, before falling back in torrents. The sky ignited into fire as pebbles of rock fell back down on re-entry, billions at once. For hours, the horizon blazed continuously, the world beneath scoured by what returned via gravity.

Forests across entire continents ignited in an instant. Flames tore through the canopies. Trunks turned into pillars of charcoal. Anything within the immediate blast radius would have been incinerated instantly. Herds of hadrosaurs stampeded and collapsed. Their escape was futile; lungs seared by each scalding breath. Above, pterosaurs folded mid-flight. Their wings curled as their bodies turned to soot, drifting down like black confetti. The silhouettes of trees and animals were burned into stone. Scorched outlines were all that remained.

Then the fire gave way to ash.

Dust, soot, and sulphates surged upward, spreading until the Sun was hidden; day suffocated into night. Photosynthesis ceased. The seas soured—their surfaces, once blue and green, turned grey and brown; their depths stripped of light. Plankton starved first, then fish, then coral reefs, and finally everything that depended on these food chains.

Seventy-five per cent of all species vanished in the aftermath of this collision—one of the most violent days in the history of life on Earth.

Even as the fire faded and the ash settled, twenty-five per cent of life clung on. Beneath the soil, in burrows and sheltered shadows, the small and the resilient held on. The world that had seemed finished was broken open; the age of giants ended, and in the silence that followed, a new story prepared to begin.

From the Ashes

In the aftermath of the asteroid's impact, survival belonged to the resilient few. Crocodiles slid into murky waters; turtles folded beneath mud flats; snakes and lizards coiled into cracks and lay unmoving. In pockets sheltered from the worst of the devastation, mammals held on too. They could vanish into the earth or foliage and wait.

When the sky collapsed and forests burned to ash, traits that once held mammals back ceased to be limitations. They became strengths that offered opportunity. From that apocalyptic ending, mammals outlasted the nightmare long enough to inherit the world.

For decades, the skies hung heavy with soot; it would have resembled a global nuclear winter for centuries. Tens of thousands of years passed in limbo, caught between death and rebirth. Then eventually, light filtered through. The haze dispersed, and the Sun returned in a dull glow, eventually becoming brilliant once more.

The first green came as ferns. Their spores rode the wind, settling in the cracks of ash. Within years, they unfurled fronds across the valleys. This fern spike, visible in fossil pollen, is the signature of renewal. What had been deserts of charcoal became carpets of green. Insects followed. Beetles bored into bark, flies swarmed over rotting corpses, and larvae fed in the damp remains.

Life would have been challenging for anything that survived. Imagine a shrew-sized ancestor squeezed into a burrow beneath a toppled tree. The air above remains heavy, making it difficult to breathe, but underground, her young cling, blind, to her belly. Each night she ventures out, whiskers twitching, claws sifting through ash for beetle grubs or seeds the flames left behind. Every successful return to her burrow is a triumph measured in mouthfuls, in the trickles of milk her body can spare.

Multiply her tale across countless burrows, and you glimpse how mammals crossed the bottleneck: litters survived, insects were scraped from ash-choked soil, and countless creatures took refuge deep enough to outlast the cold.

As time loosened catastrophe's grip, the mammals expanded outward into the new ecological openings. Some remained small, hedging survival on secrecy; others stretched into new shapes, testing the empty niches left vacant. They bred rapidly, lived briefly, and adapted constantly. In a ruined world, being a generalist was the winning strategy. Each successful generation carried forward a possibility much larger than any one animal could have grasped.

The Dawn Of Mammals

Ten million years had passed since the asteroid stripped the planet bare. The skies cleared, and the Earth was once again draped in green. Forests covered the continents–ferns, palms, and flowering trees pushing into ground once scorched to ash.

Into this renewed tapestry, mammals already diversified beyond their ancient margins. Some were still small, but others grew bold. *Pantolambda*, about the size of a sheep, lumbered through the undergrowth on padded feet. Its snout rooted among the ferns for leaves and shoots. By rivers, *Coryphodon*—a blunt-toothed, semi-aquatic browser the size of a small rhinoceros—waded belly-deep in the shallows. It tore at water plants with clumsy jaws.

Creodonts–ancestors of no modern carnivores–patrolled new ecosystems. Their jaws held the bite force of future big cats, even as their long, low bodies still hinted at a heritage from a reptilian world. They stalked small ground sloths no larger than dogs, or early hoofed ungulates that scattered through the clearings. Hunts were messy; the balance between hunter and hunted was not yet as refined as in later periods.

In the canopy, the earliest primates were just beginning to find their branches. *Purgatorius*, smaller than a rat, gripped the trees searching for fruit. Its eyes were set forward in a way that would, millions of years later, hold the same orientation as our own. From such humble experiments would come the great apes, but in this moment, it was just another small survivor.

Seen retrospectively, these new expressions of evolution look like inevitabilities on a path toward the future. Lived in their moment, they were anything but inevitable. A forest filled with rustling mammals was still a fragile improvisation, easily undone by famine, climate swings, or the slow hand of chance. Despite this precariousness, they grazed in the clearings, climbed in the canopy, and hunted in the undergrowth.

The fall of the dinosaurs left a vacancy so vast it would shape every lineage that followed. Their disappearance became the very frame for evolution itself, as decisive in its power as their reign had once been. The mammals progressed because the stage was open for them. The age of reptiles ended in disaster, yet their imprint bleeds into the present. Without minds capable of remembering them, the dinosaurs would have been nothing more than fossilised bone.

But human imagination stitched their remains into bodies, and then stories of their lives, and carried their image far beyond the strata that hold their fossils. Dinosaurs survive as a cultural juggernaut—larger in our collective memory than many other extinct animals closer to us, even those that lived 66 million years ago. Through our fascination with them, their memory lingers as vividly as any living beast.

For all the knowledge and power of modern humans, it almost defies belief that we are descended from such humble, hidden creatures. If history had turned but a fraction differently–if the asteroid had missed, or one more creature had failed to find shelter–we might never have even existed. By such a slender thread of survival, an entirely new world was born. And from the ashes of one world sprang our story: the improbable tale of mammals who dared to endure, carrying the promise of us in their tiny, beating hearts.

16

Dragons in the Dark

Mesozoic—Triassic–Jurassic–Cretaceous (≈252–66 million years ago)

Dinosaurs are the most famous animals never seen alive. Their bones, pulled from stone just two centuries ago, made them part of human culture. They began in museums and books, then leapt into films and toys. Their names are spoken by children who learn to pronounce *Parasaurolophus* before they know their own address. I know, because I was that child.

It's easy to see why we find these animals so fascinating. Stand beneath the ribs of a sauropod, and the body has the presence of a building. The arches sweep overhead like the vault of a cathedral. The spine and tail stretch away until they vanish into the distance. Looking upwards, one can see how life expanded to sizes that our minds can't comprehend.

Then you look at the skull of a tyrannosaur, a metre and a half in length. The musculature that anchored to its ridged skull gave it a bite force measured in tonnes. The appearance of a fossilised skull alone is enough to force us to confront an anatomy designed to dominate prehistoric ecosystems. Set beside them, even the apex predators of our own age—tigers, jaguars, and bears—feel diminished, their ferocity rendered provincial. The difference feels as stark as a saltwater crocodile confronting a house cat.

That dual register goes a long way toward explaining their resonance with our species. Dinosaurs are real, shaped by pressures as physical as those that honed reptiles and robins. Simultaneously, they inhabit the imaginative territory of dragons and monsters. The only difference is that these monsters left bones behind.

But myth distorts as much as it preserves. Dinosaurs were presented theatrically from the start. Nineteenth-century palaeontology was egotistical rivalry under the guise of science: *Megalosaurus*, *Iguanodon* and *Brontosaurus* were discovered, extracted and unveiled as fast as possible. Unsurprisingly, they were mounted in error: *Iguanodon* with its spike as a horn, and *Brontosaurus* with its tail dragging slack across the ground.

Each generation has remade them as we pull the story together. For Victorians, they were gothic relics of a vanished creation. In the anxious mid-twentieth century, they became parables of failure, lumbering reptiles unable to adapt, cautionary tales to empires that feared their own collapse. Science eventually linked raptors to birds after finding feathers imprinted in stone; chemistry revealed pigments; and biomechanics allowed us to predict their speed. This gave us a view of these animals that aligned much closer with reality.

Hollywood seized the science and amplified it. Dinosaurs were irresistible; colossal enough to fill a screen, alien enough to astonish, and free for anyone to use. No studio owned them, yet they carried the cultural weight of the most valuable modern movie franchises. Their bones were already on display in museums, their names embedded in classrooms, their silhouettes etched into our collective childhoods. They were a mythology shared by many, ready for retelling, and humans were more than happy to listen, repeatedly.

Dinosaurs absorb these projections because of the peculiar space they hold in the human mind. They are unbelievably remote, yet their bones are tangible; mammoths feel familiar, hairy elephants that *Homo sapiens* have encountered before. Anything before the dinosaurs feels too estranged and abstract for mainstream appeal. Dinosaurs sit in the middle, close enough to imagine alive but strange enough to induce awe.

Dinosaurs linger in the imagination because they seem unbound by physical scale and unusual appearances. However, their true dominion stretched across deep time as well. While our species has chipped tools for a few hundred thousand years, dinosaurs lived through the opening of oceans, the tearing of continents, and the slow drift of the planet itself.

The Deep Time of the Mesozoic

The reign of the dinosaurs stretched for 186 million years. We call this the Mesozoic age. Speaking the name is easy; grasping the time it represents is more difficult. Our chronicles and monuments barely touch that timescale. Measured against the age of dinosaurs, the whole of human history folds into a margin too narrow to even identify.

Every orbit of the Earth carries the solar system further through the galaxy. By the time the last dinosaur fell, our planet had drifted thousands of light-years from the place where the first had hatched. The arc of their dynasty spans both continents, tearing themselves apart, and the very repositioning of Earth within the Milky Way. Geologists break the Mesozoic into three sections, using extinctions, upheavals, and rebirths in the strata to make sense of the immensity.

The Triassic (252–201 million years ago) began in the aftermath of another mass extinction our planet witnessed. Life had been scoured by the Permian extinction. The Triassic was unstable, a planet clawing its way back into complexity in an unforgiving world.

The Jurassic era (201–145 million years ago) widened the scope. Forests spread across the drifting continents, conifers rising into lofty canopies. Sauropods joined the heights, their necks arcing above trees, their weight pressing valleys into plains. Predators, arms shortened into hooks, skulls and jaws perfected into weapons. In the same span, the first birds drew air into their hollow bones and tested the sky.

The Cretaceous (145–66 million years ago) brought with it a climax. Flowering plants unfurled across the continents, changing the texture of light and colour, drawing insects into new contracts of pollination. Horned herds and duck-billed grazers filled floodplains in numbers the Earth had never seen. *Tyrannosaurs* perfected their predatory skills, becoming among the most successful carnivores of all time. By the end of the Cretaceous, ecosystems were layered, intricate, and crowded with invention.

Dinosaurs reigned across spans so vast that entire mountain ranges rose and wore down while their lineages continued. Oceans opened and closed, climates rewrote themselves, plants remade the chemistry of the air—and still dinosaurs thrived. Confronting the expanse of their time is to feel our own brevity exposed. In the long-term game of survival, life's greatest achievements are measured by the ability to persist through the churn of deep time.

And even they ended. No span, however vast, is immune to catastrophe. The Mesozoic lasted almost two hundred million years, yet it still closed in fire. What unsettles us is that the day of their ending dawned like any other, and nothing prevents such a day from coming again.

Monsters of Our Own Making

Dinosaurs loom so large in the imagination that it is tempting to see their world as a parade of giants, a time when every animal was a fortress of teeth and claw. The names we remember are the celebrities: *Tyrannosaurus*, *Brachiosaurus*, *Stegosaurus*, *Velociraptor*, *Triceratops*. The first Jurassic Park film embedded this vision so deeply that an entire generation could picture them side by side, as if they had actually once shared the same island. It feels coherent and fits the rhythm of storytelling for entertainment. But in truth, many of these animals were divided by tens of millions of years. *Tyrannosaurus* lived closer to the age of smartphones than it did to *Stegosaurus*. The familiar image of a Cretaceous floodplain where every icon gathers together is more a projection of our present than an accurate telling of the past.

The distortions run deeper than chronology. We imagine their age as an age of monsters, when in fact dinosaurs filled the world in far more diverse ways. They were often creatures the size of deer, dogs, or turkeys, living lives as ordinary as the animals around us today. They hunted insects in the undergrowth, browsed ferns in the shade, and raised broods in nests not unlike those of modern birds. The giants—the sauropods, the theropods—were mostly outliers, their combined weight only a sliver of the planet's living mass.

The dynasty was sprawling, intricate, and diverse enough to fill every ecological niche. Based on what the fossil record allows us to reconstruct, palaeontologists estimate that on the order of a few thousand dinosaur species lived during the Mesozoic, implying billions to perhaps tens of billions of individual animals, while recognising that the overwhelming incompleteness of the record means the true total may have been orders of magnitude larger, beyond anything we can presently count with confidence. Most of them were forever erased. In rock, only the slightest evidence remains.

The reason for the lack of remains lies in the nature of preservation. Fossilisation is not the natural fate of every corpse but a rare convergence of chance. A body must be buried promptly, starved of oxygen, shielded from erosion, and then survive for millions of years as the earth shifts. What remains is often partial: a tooth, a fragment of jaw, a few scattered vertebrae. Even the great *Tyrannosaurus rex*, among the most intensively studied of all dinosaurs, is represented by fewer than fifty skeletons in anything approaching completeness. From scraps like these, entire worlds are reconstructed. The result is fragmentary—a catalogue in which careful science must be used. Thankfully, human imagination and problem-solving allow us to lean on science to fill the gaps and imagine these animals alive.

The Dinosaurs of Tomorrow

The asteroid tore down one dominion and cleared the ground for another.

Dinosaurs flourished. They radiated into thousands of forms, adapted to forests, deserts, swamps, and coasts. If survival is the test, they passed it with distinction.

For all their merits, across all those millions of years, something was missing. Dinosaurs mastered speed, size, armour and feathers. But they likely did not approach the symbolic intelligence that would later define mammals. Their brains stayed within reptilian limits. Some groups—troodontids, dromaeosaurs—pushed the envelope, carrying brains relatively large for their size. They may have shown complex social behaviours. But no evidence suggests they crossed into abstraction, into imagination, into the kind of thought that could one day dream itself into history.

Over one hundred and fifty million years of dominance, and intelligence—in the way we think of it—never arrived.

Why? Because of an angle, a trajectory, a strike zone no wider than chance itself?

Only thirteen per cent of the Earth's surface holds the chemical cocktail needed to turn an asteroid strike into an apocalypse: limestone, gypsum, carbon-rich rock, materials that, when vaporised, unleash sulphate and carbon dioxide enough to choke the sky. Chicxulub was one of those rare locations. The asteroid approached from the Northeast, the angle of entry shallow enough to maximise the debris ejected into the atmosphere. Had it come hours earlier, it would have hit deep water that absorbed much of its own fury. Hours later, and it might have struck continental crust with less sulphate to release. A different tilt may have diminished the fallout. Dinosaurs might have staggered, but recovered.

Instead, the dice landed on the unlikeliest face. The exact place, the exact angle, the exact geology. Enough to turn stability into cataclysm, to erase entire lineages over a few months, an ecological reset.

The mammals, hiding in burrows and shadows, inherited by default.

But the counterfactual lingers: had the asteroid missed, what then?

Birds—the living dinosaurs—prove that their lineage was not destined for decline. Many non-avian forms were diversifying right to the end. Some were experimenting with behaviours more intricate than we once thought possible. Give them another fifty million years—the span of time it took primates to move from tree-dwellers to tool-makers—and they might have stepped onto a different trajectory altogether.

The idea that intelligence must be mammalian is not inevitable. Just as likely, it could have been reptilian. It could have worn feathers. It could have hatched from eggs and hissed through tooth-lined jaws. But intelligence is not a straight road, as evolution does not strive for it. It is directed only towards survival and reproduction. For most of life's history, being "smart" was irrelevant. Bacteria thrive without thought; ferns conquered continents without neurons. Trilobites endured for 250 million years with nothing more than crystal eyes and simple instincts.

So if dinosaurs had continued, would natural selection ever have rewarded them with the wiring needed for culture, art, or technology? Perhaps, or perhaps not. It is likely that their world would have remained primitive by our standards, honed senses and refined behaviours, without ever needing symbolic imagination. Perhaps they would have remained perfect at what they already were: predators and grazers locked in an eternal arms race, endlessly successful, mindless by our standards.

Had the asteroid missed, could any lineage of feathered reptile have inched beyond mere instinct—could they have learned to guard a flame, to weave memory into story, to build a civilisation whose ruins might one day puzzle a mammal that never came to be? Would language or hierarchies ever have taken shape? Might an advanced reptilian world have drawn borders, raised militia and crowned a powerful few to rule the many?

When our own age came around, we proved no wiser.

The boundary between what happened and what almost did is thinner than we like to believe. A fraction of a degree, a few hours of planetary rotation, and the world could still belong to the descendants of dinosaurs.

We will never know. The impact at Chicxulub collapsed those futures into silence.

That vacancy was our chance.

And chance defined everything.

The asteroid collapsed futures we will never know. But what we do know is that the rule of the dinosaurs was a sequence of movements, each broken by upheaval and renewal. The Earth itself preserves that order in fossilised remains and in the genes of its descendants, its strata dividing the Mesozoic into those three great acts. To follow those layers in detail is to see their dynasty in its true scale—a history of life written across the churn of deep time.

17

Thrones of Tyrants, Crowns of Horns

Cretaceous (≈145–66 million years ago)

The Cretaceous marked one of Earth's most extravagant eras of deep time. Seas rose more than 200 metres above today's coastlines, submerging familiar lands beneath warm, shallow waters. Near the equator, intense heat fueled violent storms, with massive clouds building and collapsing in torrents of lightning and rain. Forests of magnolia, sycamore, and early figs emerged, their colours contrasting with ferns and cycads. The world was a lush greenhouse, brimming with life yet teetering on the brink of excess.

Continents had broken from their supercontinental past and drifted towards more recognisable outlines. North America was split in two by a shallow ocean running from the Gulf of Mexico to the Arctic. Europe was scattered into archipelagos, and Asia broadened towards its immense future mass.

During the Cretaceous, dinosaurs occupied nearly every terrestrial ecosystem. Yet, this age offered more than a stage for reptilian splendour. Flowers began to rise. Their colours and perfumes drew insects into new relationships, weaving alliances that still shape our diets and landscapes. Seas swarmed with mosasaurs and plesiosaurs. Ammonites drifted in spirals of shell. Sharks edged toward their modern forms. In the skies,

Quetzalcoatlus tested the limits of biomechanical flight.

Mammals remained in the margins. They gnawed seeds, nested in burrows, and carried young in pouches or wombs. These creatures were experiments, still constrained by the weight of dinosaur dominion, rehearsing strategies that would only matter once the stage was cleared.

Dinosaurs at the Cretaceous Apex

The regime of the dinosaurs, already ancient, reached a pitch of diversity and scale. It reshaped the very fabric of the Earth. Across floodplains of North America, herds of hadrosaurs migrated in vast numbers. Their passage carved paths into the land. They trampled vegetation into mulch, fertilised the ground with waste and drove forests to regrow. Their grazing pressure created open clearings. Mammals, birds, and insects found space to flourish there. Horned ceratopsians used their armoured skulls in conflicts. Their frills and horns clashed in displays of aggression and protection.

Raptors stalked the forest margins, their feathers bristling, their vision locked forward in binocular focus. While still debated, they likely hunted as coordinated, intelligent predators. And at the top of the food chain, *Tyrannosaurus rex* ruled the northern continents for roughly two million years.

Dinosaurs had been present since the Triassic. In this age, they reached their fullest range. They spilt across continents, filled the skies and waded into shallows. While their bodies carved the land, another revolution was at their feet. Its strategy lay in subtler forms: petals, scents and seeds. These would carry forward into worlds the giants would never see.

The Flowering Revolution

A single bloom might seem inconsequential against the weight of a dinosaur. Picture a hadrosaur pressing through a floodplain, its bulk shaking the ground, its mouth filled with ferns and conifers. In the shadow of its stride, a flower opens: a burst of colour, a trace of scent, a lure for insects.

Until this point, Earth's greenery had relied largely on wind, water, and probability. Ferns shed spores. Conifers loosed clouds of pollen. Cycads cast their dust on the wind. Reproduction was a lottery, cast into the air or water without certainty. Angiosperms changed the rules by forging transactions: colour and nectar for transport, scent and shape for precise locating. Bees, beetles, butterflies and flies became partners in this flowering bargain. They carried pollen with an intent wind could never match. Each visit completed a circuit and multiplied abundance.

Dinosaurs grazed in worlds where food no longer arrived at a flat, steady pace but in crescendos—floods of fruit and foliage, cascades of colour that demanded attention and adaptation. Herbivores with jaws and guts suited to this new bounty flourished. Insects diversified into a dazzling variety, tracking the blossoms' invention with their own adaptations.

Early angiosperm (the scientific term for flowering plants) debris washed into rivers and estuaries. This altered nutrient flows and reshaped coastlines already transformed by the swelling of the Western Interior Seaway. New sediments were buried and preserved, setting the stage for the fossils we now unearth. On land and in the sea, the flowering surge re-engineered environments as surely as tectonics or storms.

By the time the asteroid fell, the flowers—only tens of millions of years into their story—were already secure. They survived the destruction that followed and multiplied as the Palaeogene dawned. Angiosperms transcended the Mesozoic and became part of the story of mammals, primates and ourselves.

Shallow Oceans With Great Wings Above

If you could stand on the shoreline of the Cretaceous Interior Seaway, the horizon would be nothing but water and islands. Warm tides lap against mudflats where reeds and cycads root in brackish soil. Dragonflies hover in the dense air, their wings catching the sun. Overhead, the pterosaurs drift across the shallows, their wingspans casting giant shadows.

The water darkens out beyond the tideline. A head breaks the surface: a mosasaur. It bursts from the depths with a muscular roll, its jaws hinged wide to take a struggling ammonite. Then it sinks again, leaving only ripples. Often mistaken for dinosaurs, these predators were in fact lizards that had returned to the sea, descended from ancestors that once walked on land. Their bodies lengthened into muscular torpedoes, their limbs remodelled into paddles, their tails like rudders. They were predators that seized an evolutionary opportunity in Cretaceous waters.

Some species of Mosasaurus reached lengths of 15 metres. Fossilised stomach contents show they ate fish, ammonites, diving birds and even other mosasaurs. The Cretaceous was dangerous beneath the surface: plesiosaurs cruised with spear-like necks, sharks refined their already sleek designs, and ammonites filled the water column in their millions. Yet the mosasaurs imposed themselves as the apex. Their reign was brief in geological terms—thirty million years—but in that span they forged the last great marine empire before the asteroid erased it.

Above them, pterosaurs had mastered flight for over a hundred million years. They still launched from cliffs and estuaries. Some grew to stupendous size. *Quetzalcoatlus* had wingspans rivalling a small aircraft, gliding across thermal currents with the grace of kites but the tenacity of fighter jets. Their bones were hollow but reinforced. Their frames were both light and strong.

In the same skies, birds were gathering strength. Small feathered dinosaurs had already tested gliding in the Jurassic, but in the Cretaceous, their descendants refined true flight. *Archaeornithura* and *Ichthyornis* darted over rivers, their wings feathered, their brains expanding in capacity for coordination and memory. For a while, both lineages shared the air: leather-winged pterosaurs above, feathered newcomers below. Designs clashed, and though pterosaurs thrived, birds would be the ones to survive the Mesozoic's collapse.

Dinosaurs ruled the land, yes, but the oceans and skies carried dynasties just as extravagant. Coral reefs flourished into architectures not so different from modern ones, crowded with ammonites whose spirals still mark fossils on our museum shelves. Fish radiated into shapes that remain familiar, from tuna-like hunters to ray-finned grazers.

Abundance on the Brink

Under the Cretaceous skies, carbon dioxide rose to levels four—perhaps even ten—times pre-industrial concentrations, creating an atmosphere that would feel oppressive and stifling to us now. Forests stretched into latitudes that in our age lie locked in snow. Ginkgo and magnolia grew in Arctic light that lingered as twilight for months, feeding herds of dinosaurs that walked beneath a sun that never truly rose.

Nearer the equator, heat gathered into storm towers. These climbed kilometres into the sky. Lightning raked the horizons. Rains poured until rivers tore fresh paths across floodplains.

This was a greenhouse Earth, and it came with excess. Plants grew in surges, their canopies expanding in thick, humid waves. Fire followed close behind, ignited by lightning and fed by dense undergrowth, only to enrich the soil and restart the cycle. In the oceans, heat pushed chemistry to the edge. As water warmed, oxygen thinned, and vast dead zones spread, where nothing larger than microbes could survive. When plankton bloomed and collapsed in their billions, their remains settled into bands of chalk that now rise as white cliffs in southern England—a Cretaceous sea recorded in stone.

That is why the Cretaceous speaks with two voices: triumph—the apex of a dynasty that reshaped every landscape it touched—and prelude, a reminder that abundance and collapse are born of the same forces.

From here, the path runs backwards, into the Jurassic. The first great planet-spanning forests emerged there, sauropods elongated their necks, and the first birds flapped their wings.

18

The Ascent of Giants

Jurassic (≈201–145 million years ago)

The word Jurassic carries a double weight. To the world at large, it belongs to cinema, a shorthand for spectacle and for our own tentative meddling with life's ancient code. Fixed by Michael Crichton and the films his story inspired, the name conjures predators that never walked its forests, faces borrowed from other ages, stitched into monsters. Yet long before Hollywood, the word came from stone. Early geologists first claimed it from the Jura Mountains on the French–Swiss border, where limestone cliffs held the bones and shells of creatures from an age no human had imagined.

Pangaea, the last supercontinent, cracked along the seams and drifted towards the forms we know. Oceans poured into the split, climates fractured, and populations that had once moved freely across a single landmass became isolated, setting evolution onto new divergent paths. The Jurassic was the long expansion between two cataclysms.

The Masters of Magnitude

Nothing represents the spectrum of life's diversity in the Jurassic more than the sauropods. Necks the length of bowling lanes, frames heavier than airliners, hearts forcing blood 10 metres skyward. Brachiosaurus may have stretched 30 metres, stood 7 metres at the shoulder, and pressed the ground with the weight of 60 tonnes. These were the largest land creatures the Earth has ever carried.

Their vertebrae were hollowed with air. This lattice supported necks fifteen metres long that did not collapse. The same system of sacs threaded through the skeleton was tied to lungs that breathed in a single direction, drawing air as efficiently as birds. Evidence now shows that those sacs extended even into the tail, so the body became light through the use of air.

Their heads were usually tiny when compared to the rest of their body. They snipped leaves, sending them to stomachs where fermentation vats processed tonnes of vegetation. Stones were swallowed with the intention of helping grind up tougher plant matter, but microbes did most of the work. The world they lived in provided the fuel these gargantuan animals needed. Jurassic forests, swollen with carbon dioxide, grew at a pace that matched their appetite.

Beginning as hatchlings no larger than adult chickens, they reached adulthood in just over twenty years. Their eggs were small, rarely larger than a football. Physics prevented much more; a bigger egg would suffocate its embryo. Dozens were laid in each clutch, enough to ensure survival by chance. Once full-grown, they were almost unassailable. Predators such as Allosaurus prowled their world, but a sauropod tail could break bones and scatter hunters. Size is the ultimate advantage in such a dangerous landscape.

What remains of them is fragmentary; only several complete skeletons have ever been discovered. Their bones, lightened by air, shattered easily, leaving femurs taller than men scattered across continents. From these pieces, we rebuild the outline of what once walked here, stone puzzles that talk to a scale no living land animal approaches.

Only the oceans have nurtured larger bodies. Blue whales exceed even the largest sauropods in mass, but their weight is lifted by the embrace of water. Gravity is softened there; buoyancy solves the problem that land could not. They drift through seas made rich in plankton blooms, sieving swarms of krill by the tonne, nursing calves that enter the world already larger than cars. On land, however, sauropods remain without rival everywhere else in the fossil record.

Beyond Earth, other answers may exist. On worlds where gravity pulls more lightly, or where atmospheres allow for it. If biology creates complex life elsewhere, it may yet build forms larger still. One can imagine forests under red suns where bodies tower beyond our reckoning, creatures engineered by parameters nothing like our own.

But here, within the narrow tolerances of this planet—under this gravity, in this Jurassic air—sauropods marked the outer edge. They were the greatest terrestrial animals that ever lived, an evolutionary audacity never repeated.

A Splitting World

At the dawn of the Jurassic, the Earth was bound by a single landmass. Pangaea stretched almost from pole to pole, a continent so vast that its centre lay starved of rain. Deserts covered its interior, unending plains of dust and outcrop, while forests clung to the margins where ocean winds still carried moisture inland. For tens of millions of years, this supercontinent had unified the world, funnelling animals across its breadth, holding climates in a single grip. Beneath its surface, though, the planet was restless. Plates strained, pressure built, and the crust tore from the middle.

The rift that opened between what would become North America and Africa was an event of creeping violence. Valleys split apart, their walls striped with fire, their floors drowned in basalt. Lava spilled in volumes, blanketing the land in black floods that cooled into scars still visible in Morocco, Brazil, and Nova Scotia. In time, the fissures widened and admitted the sea. Saltwater crept into the wounds of the continent, and the first waters that would become the Atlantic glistened in the Jurassic sun.

With that division, the climate shifted. A single landmass forces uniformity; its interiors dry, its extremes harden, and its weather stagnates. When the continents fractured, coastlines multiplied, winds changed, and the monsoons gathered strength. Rains swept inland, rivers braided across floodplains, and forests pushed forward into valleys that had baked for millions of years. The Jurassic was warm, its skies rich in carbon dioxide, and plants answered with furious growth. Conifers rose across hillsides, cycads crowded the clearings, ginkgos stood in groves, and ferns laced the ground into carpets. Before the rise of flowering plants, these older lineages thrived under the Jurassic conditions, multiplying into forests and meadows that fed giants and transformed entire landscapes.

Dinosaurs once roamed across Pangaea without constraint, but now found themselves separated. Some populations diverged in isolation; others poured into new green worlds and adapted at haste. Diversity accelerated to match the new environmental parameters, and the Jurassic cast assembled.

In Laurasia, one of the continents that diverged from Pangea, Allosaurus thrived, its skull lightened by hollows like those of the sauropods. Armour rose in answer to new predators. Stegosaurs carried rows of plates along their spines and spikes at their tails. Their plates may have flushed with colour in contests of display or shone pale in the heat of the sun. The spikes could break bones. Ornithopods grazed in groups across the floodplains, their survival tied to numbers and escape. Coelurosaurs, feathered and light-boned, darted among the ferns.

The oceans, too, bore the marks of change, inviting ichthyosaurs and plesiosaurs into new territory. Shallow seas spread along the coasts, multiplying habitats and drawing marine reptiles into niches untested before.

By the middle of the Jurassic era, the Earth was fractured and fertile. Deserts had retreated, forests and rivers stretched where interiors were barren, and coastlines wound like lace around the new, separated worlds.

The fracture of Pangaea, and the path life took through the Jurassic, lay bare a principle as old as life on Earth. Living things are never sovereign. They move in step with the slow machinery of the planet: with climates that open and close like doors, with continents that grind together and tear apart. Forests, herds, predators, even the giants themselves—each reflected those shifting conditions, able to thrive only because the Earth allowed it.

The Skies of the Jurassic

The Jurassic sky was teeming with life. Pterosaurs, masters of the air since the Triassic, glided over seas and floodplains, but toward the close of the Jurassic, one of the defining discoveries of this line remained preserved enough to tell us everything. It was tiny, half-reptile and half-bird. The transition we had been guessing was caught in progress.

Around 150–148 million years ago, in a limestone lagoon in what is now southern Germany, a feathered dinosaur sank into lime mud, its body sealed in layers so fine that each feather quill left its imprint. When quarrymen split the slabs in 1874, a snapshot into the void emerged: teeth in the jaws, claws on the wings, a long bony tail—and feathers fanned into vanes fit for flight. This was the discovery of the Berlin Specimen of *Archaeopteryx lithographica*, a feathered dinosaur with teeth and a tail, preserved at the threshold of the transition from reptiles to birds.

A crow-sized creature with the skeleton of a reptile and the plumage of a bird, as if nature had been caught in mid-sentence. Its wrist bore a pulley joint that let the wing fold tightly, protecting the flight feathers. Its breast carried a wishbone that flexed with each stroke, storing and releasing energy like a spring. These were some of the first outlines of a body that could lift itself into the air.

The flight it managed was likely short and uncertain—bursts from branch to branch, glides across lagoons, quick escapes into the canopy. But it was a true flight all the same, the first powered ascent of a dinosaur into the air. Air sacs lightened its bones and fuelled its muscles. Even its brain had shifted, expanding the regions for balance, vision, and coordination—changes demanded by three-dimensional movement.

The lineage that produced birds had been shrinking for millions of years, bodies becoming lighter, generations turning faster, mutations spreading with speed. In a world that celebrated size in sauropods, evolution had written another lesson: to be small was to be quick, and to be quick was to adapt.

At first, these new fliers held little presence in the sky. Pterosaurs still soared higher and further. Birds were marginal, seemingly inconsequential. If the story had ended with Archaeopteryx, it would be a curiosity. But the line survived, and in surviving, it transformed. Teeth gave way to beaks—light, sharp, and highly versatile. Diets diversified, from insects to seeds to fruit. By the Cretaceous, birds were branching into dozens of forms, filling forests, riversides, and coasts. When catastrophe struck, this flexibility was essential. Seeds buried in soil, waiting out the dark, kept them alive when leaves vanished, and giants starved.

The syrinx, a vocal organ deep in the chest, does not appear in the fossil record until the end of the Cretaceous. Jurassic skies would have been quieter, save for the rush of wings and the call of pterosaurs. Only later did song accompany flight.

The Jurassic perfected magnitude on land and splendour in the air. It gave us sauropods the size of buildings and pterosaurs whose wings traced the horizon. But the survivors, the heirs, were these crow-sized fugitives experimenting with feathers. Beneath the spectacle of giants, the lives of mammals and the first birds carried the undercurrents of a future they would one day inherit.

The absurdity resolves into inevitability. Beneath the scale of the Jurassic, other threads crept, their presence obscured by the splendour of the age. The Triassic lies next, stranger and far less forgiving than the Jurassic: a world of turbulence where life clung on against the odds.

19

Interregnum of the Archosaurs

Triassic (≈252–201 million years ago)

The Triassic was the most alien of the Mesozoic ages. Its horizons stretched across the single supercontinent, Pangaea, not yet broken apart. Along the margins of this land, monsoons hammered the coasts, driving torrents of water into valleys that lay parched for years at a time. The air itself was heavy with carbon dioxide, thick enough to draw forests towards the poles and feed wildfires that raced across tinder-dry plains each season.

Oceans lay sluggish, often depleted of oxygen, while on land, the survivors gathered where they could. Fossil beds show the same life forms recurring. Ecosystems rose, faltered, and fell silent. Adaptations that should have secured an advantage were swept away by drought or flood. Life marched forward only to be beaten back, as if the planet itself refused to provide stability.

Within this turbulence, amphibians with skulls as long as a man was tall lurked in stagnant rivers, waiting like living traps. Armoured reptiles rooted through floodplains, bulldozing ferns and cycads aside. Delicate gliders leapt from the trunks of ancient conifers, steering with wings stretched between their hind limbs. Among them, small archosaurs refined the traits that would one day dominate the Earth. From them came the first true dinosaurs, fox-sized hunters scarcely noticed in their time. Above, the first pioneers of flight, early forms of pterosaurs that traced new paths through the sky.

The earliest Triassic landscapes bear the record of this uncertainty. Fossil beds across South Africa's Karoo Basin, Russia's plains, and China's river valleys reveal the same picture repeated: a handful of hardy species clinging on in places where entire ecosystems should have stood. These "disaster faunas" were sparse, provisional gatherings of survivors. It was, in the old Roman sense, an interregnum: the interval between rulers when no crown was secure. Here, the archosaurs held that gap, waiting for the moment when their lineage would seize the throne of the future Mesozoic.

Communities expanded, only to vanish; new species appeared, only to die out again. The Smithian–Spathian crisis, a sharp climatic upheaval within the Early Triassic, knocked life back just as it seemed ready to regain balance. Layer by layer, the strata show turnovers of species so abrupt they read like continuous false starts—evolution pressing forward, the environment striking back.

This drawn-out struggle makes the Triassic unlike the other two periods of the Mesozoic age. Adaptations that in later eras would have secured dominance—burrowing, armour, specialised jaws—were tested here and broken against the volatility of the planet. It was a world where survival meant enduring through collapse, where resilience mattered more than innovation, and where even the hardiest creatures were passengers of a climate in turmoil.

The Sovereign Herbivore

If the Early Triassic had a sovereign, it was a herbivore: *Lystrosaurus*—squat, tusked, with a beaked snout and limbs that sprawled from its body. For a time, it became the most abundant large land animal the planet has ever known. Its fossils overwhelm the strata of the age. In South Africa's Karoo, in Russia's floodplains, across India and China, even in the polar forests of Antarctica, the record repeats the same refrain. In some beds, nine skeletons out of ten are *Lystrosaurus*. Imagine an Earth where almost every creature you saw bore the same shape.

Barrel-chested and low to the ground, it moved with the awkward gait of an animal neither reptile nor mammal. Its head was capped with a keratinous beak for cropping plants, tusks jutting forward like chisels. Its skin was likely thick, its appearance ungainly; this was the body that carried ecosystems through a fractured world.

Lystrosaurus could graze on whatever foliage clung to riverbanks, burrow into soil when the heat pushed down, and wallow in mud to cool its body. Its tusks preserve rings that faltered in drought. But this was a strange dominance. Most worlds divide power among many species, predators and prey bound in complex webs. The Early Triassic narrowed it to one. A single herbivore filled the vacuum, its body the scaffolding of ecosystems too fragile to sustain variety—a ruler by default in the silence left by extinction.

However, its existence was significant. By holding space, by continuing to feed and reproduce across a shattered biosphere, *Lystrosaurus* gave time for recovery to begin. Plants spread in its wake. Insects multiplied, and new vertebrates rose around it. Life bridged the Permian abyss with Triassic experiments.

Inevitably, its age passed. As ecosystems grew richer and diversity returned, sharper and more specialised lineages overtook them. What had been for a time the planet's best survivor slipped into obscurity.

During that brief span, the Earth belonged mostly to a single herbivore. Continents stretched under a greenhouse sky, rivers lined with narrow threads of green, and everywhere this one animal survived repeatedly. The greatest monopoly ever achieved by a land animal was not the work of a predator or a giant, but of a squat synapsid. Survival often rests on persistence, and after a catastrophe, the fate of the world can hang in the humblest of forms.

Blueprints for Crocodiles and Birds

When the reign of *Lystrosaurus* ended, the Triassic continued its hostility, a world tilted towards imbalance. Life returned to this stage in distorted forms. Across floodplains, herds of rhynchosaurs gnawed cycads with beaks made for chiselling. Among them darted procolophonids, cheekbones spiked, burrowing at the first scent of heat. Pools lay in wait with amphibians of monstrous scale: *Mastodonsaurus*, its skull longer than a man. In the forests, stranger experiments: *Sharovipteryx*, kites of skin stretched from their hindlimbs, gliding from trunk to trunk.

From this fever of invention, a different lineage took hold. The archosaurs stood tall on upright limbs, their strides long, their lungs drawn through with air in a single direction. They ran where others sprawled and breathed deeper where the air thinned. Some became predators with skulls built to shear flesh and puncture bone. Others grazed or hunted in water. Their bones carried the imprint of a system still alive in birds, lungs threaded with sacs of air.

Among the Triassic experiments, it kept one design. In the archosaurs, the age found permanence until the present: a body plan that would seed crocodiles, dinosaurs, and birds, a form tuned to extremes yet able to outlast them. In a world of instability, they were the first shape to weather the storm—a law written in DNA that would govern much of the deep time ahead.

Archosaurs

The archosaurs had emerged as the victors of instability. Where the Triassic world demanded endurance and invention in equal measure, their bodies gave both. In this world, they multiplied, spreading across deserts and floodplains, rivers and forests, filling roles left vacant by extinction.

Predators of daunting scale prowled this fractured land, but they were unlike anything alive today. Rauisuchians stood high on pillar limbs, their bodies lifted clear of the ground. Their jaws snapped shut with shearing teeth that left clean cuts in bone. They moved with a gait closer to a mammal than a lizard, striding in pursuit of prey across floodplains. In rivers and lakes, their rivals were drifting just beneath the surface. Phytosaurs floated with only eyes and nostrils above the water. Their snouts were lined with interlocking teeth, perfect for seizing anything that touched the surface. They could wait for hours, then launch with an eruption of water and muscle, dragging whole animals under in sudden violence. For tens of millions of years, they were the rulers of the Triassic.

The first dinosaurs were small, fox-sized creatures in South America, darting between the legs of rauisuchians and scavenging the kills left by phytosaurs. *Eoraptor* was one of them, a quick, sharp-toothed omnivore with grasping hands and bones built for speed. *Herrerasaurus* was larger, close to two metres in length, with jaws that could shear flesh.

While the ground belonged to the rauisuchians and phytosaurs, another frontier was opening above it. Among the archosaurs, a few small lineages had begun to transform their skeletons. Fingers lengthened, bones hollowed, muscles broadened across the chest. Skin stretched into membranes that could ride the air. The first pterosaurs took to the skies. *Eudimorphodon*, scarcely a metre across, already showed the key features: an extended fourth finger that formed the leading edge of the wing, fine-boned skeletons that weighed almost nothing, and teeth like needles to snatch fish or insects. They could beat their wings to gain height, banking and swooping with control. Life had reached into the air before, with insects, but never with vertebrates.

Down in the water, another line of archosaurs was setting its course toward a different destiny. The croc-line reptiles tried many shapes—some upright hunters, others armoured grazers. Their lineage first took shape in the Late Triassic, more than 200 million years ago, among the crocodile-line archosaurs that split from the ancestors of birds and dinosaurs. From those ancient kin, the first true crocodilians emerged much later, in the Late Cretaceous, about 85 million years ago. Smaller than the surrounding giants, they kept low in the water, nostrils and eyes lifted above the surface, jaws primed with explosive power. The design scarcely wavered. From those Triassic roots to the crocodiles that slide through rivers today, this line has endured for over two hundred million years.

The Triassic-Jurassic Extinction

The Triassic was always destined to end in even more destruction. Beneath the rift that would become the Atlantic Ocean, the crust split and disgorged lava for more than a million years. Valleys drowned in molten rock, skies filled with ash, gases poured into the air in volumes no forest could absorb. Carbon dioxide spiked, heat climbed, rains turned corrosive, oceans suffocated. The planet staggered beneath its own convulsion. Rauisuchians vanished. Phytosaurs, perfected for ambush, disappeared without descendants. Some amphibians that had survived since the Carboniferous collapsed in the final blow. Whole dynasties ended, again.

Dinosaurs, once minor and overlooked, stepped into a gap cleared by this destruction. The traits that had kept them marginal—their light bones, fast growth, and ceaseless adaptability—became the keys to dominance. In the Triassic, dinosaurs were to the great archosaurs what mammals would later be to the dinosaurs themselves. When the Triassic burned, dinosaurs inherited a world emptied of rivals. When the Cretaceous ended in fire, it was the mammals' turn. In each case, the unlikely heirs rose in the next age.

The Triassic closed with this exchange: one set of rulers gone, the script handed to the dinosaurs, and through them, the rest of time was reshaped.

The Triassic began in the wreckage of the Great Dying, and it ended in the convulsions of fire that marked the boundary with the Jurassic. In between lay a world unsettled, continents joined in one supercontinent, climates pushed to extremes, and life itself still provisional. Predators rose and vanished, experiments flourished and failed, and in the margins the first dinosaurs, pterosaurs, and crocodiles continued the story. It was the long recovery in which the forms that would dominate the Mesozoic were forged, tested, and thinned. Everything made possible by the instability of these tens of millions of years.

Behind the Triassic stood an extinction deeper and darker than all the rest—the Great Dying of the Permian—a catastrophe so destructive that it almost ended life's narrative.

20

The Anatomy of Planetary Failure

Permian (c. 299–252 million years ago)

The air was stifling and bitter. Ash drifted across the horizon and settled into rivers that hissed as they boiled. Forests ignited in waves, trees collapsing into embers before rain reached them. When rain fell, it was acidic, stripping leaves and leaching soil. The oceans reeked of sulphur. Along coasts, fish floated belly-up, their gills paralysed by oxygen-starved water.

This occurred 252 million years ago.

In what is now Siberia, the crust split. For roughly two million years, in pulses, lava poured from fissures that gaped across the land. These were floods of molten rock that spread for hundreds of kilometres, layer upon layer, until a province lay beneath basalt.

The magma tore through seams of coal and other carbon-rich sediments, releasing its chemistry in volumes the air could not absorb. Carbon dioxide surged, and methane likely destabilised from frozen reservoirs on the seafloor. Sulphur and halogens rose in plumes that damaged the ozone shield and poisoned the rain. The planet had become an exhaust vent, bleeding gases that altered the climate and chemistry faster than most life could handle.

As the seas warmed, their oxygen drained away. Stagnant waters became anoxic (oxygen-deprived), then euxinic (containing hydrogen sulphide instead of oxygen). Microbes bloomed where nothing else could live, staining the water green-black, bubbling poison into the air and surf alike. Microbial mats—dense layers of microorganisms—crept back across the seafloor, reclaiming ground last held hundreds of millions of years earlier, as though the trajectory life was taking had been forced into reverse.

Even insects, among the better survivors, suffered their greatest known crisis. With plants retreating, herbivores starved. Predators deprived of prey followed. Fossil beds from this interval show silence, bones scattered without pattern, a biosphere without direction.

On the order of 80–95 per cent of marine species were lost, along with roughly 70 per cent of those on land. What remained survived by chance more than mastery: burrowers that escaped the heat, generalists that could eat almost anything. Ecosystems formed, faltered, and collapsed again as the planet lurched between extremes. The Earth seemed to have forgotten how to sustain abundance.

The most intense phase may have unfolded in as little as 20,000 years, yet recovery spanned millions of years. When stability returned, the biosphere was unrecognisable, emptied and reset. The Mesozoic Age followed because the slate had been cleared thoroughly.

To imagine this in human terms is almost impossible. An immortal observer watching over millions of years would not have seen a single moment of apocalypse but a slow erasure. Each generation would inherit thinner forests, stranger skies, emptier seas. This was a complete planetary failure, a cascade of systems breaking down together. Climate, chemistry, atmosphere, oceans, each weakened, each amplifying the next, until the machinery of life itself ground to a halt.

The Great Dying was the closest Earth has ever come to ending its story of life. It feels easy to dismiss it, the nightmare of another world. But it was this planet, under these skies, with these oceans and this air.

The Permian extinction is what happens when stresses overlap, when boundaries are crossed together, when the feedback loops that sustain life are driven into collapse. It took the planet millions of years to heal. Every flourishing you have met so far in these pages, the rise of the dinosaurs, the birds above our heads, the mammals that would one day become us, was possible only because life endured in the margins of this abyss.

The Siberian Traps

The story of the Great Dying begins in the far north of Pangaea. The crust split open, and lava poured out in pulses that each covered hundreds of thousands of square kilometres, spreading in sheets that hardened into black stone. This was flood basalt volcanism, a phenomenon that dwarfs the imagination. At its height, whole provinces drowned under flows deep enough to bury skyscrapers. When the eruption cycle ended, an area the size of Europe lay entombed beneath kilometres of basalt.

Geologists call it the Siberian Traps, from the Swedish word *trappa*, meaning stairs, a reference to the step-like basalt (a type of volcanic rock) layers laid down by repeated lava flows. But to call it a landscape feature understates its role. The Siberian Traps were one of the largest volcanic outpourings in Earth's history, and their global impact was driven by the chemistry of the gases they released and the climate changes they induced.

What happened here was a chain reaction. A feedback system that began with molten rock and ended with planetary collapse. The eruptions primed the atmosphere with greenhouse gases, the greenhouse gases warmed the oceans, the warming oceans suffocated, the suffocated oceans vented poisons into the air, and the poisons damaged the very shield that softened the Sun's ultraviolet glare. Each system broke in sequence as it shifted the biosphere's parameters.

The biosphere had endured impacts, ice ages, and even massive volcanic episodes before. But never had so many levers been pulled at once, never had so many of the planet's regulatory systems been driven into failure together. The end-Permian Great Dying, triggered by the vast eruptions of the Siberian Traps, is the clearest known case in which Earth's own volcanism, with no confirmed asteroid or comet impact, drove complex life to the brink of collapse.

And that is why it resonates with us, because the rules have not changed. When feedback loops are forced beyond their thresholds and stresses pile up faster than ecosystems can adapt, collapse is not only possible but inevitable.

Same Planet, Same Rules

We see from the Great Dying just how small the habitable zone actually is. For most planets in the galaxy, complex life never begins. Countless other worlds orbit their stars in silence, their chemistry locked in states too hostile for biology. Earth was the outlier, the improbable sanctuary. Even here, on this rare refuge, life came so close to failing. The Permian–Triassic extinction pushed the biosphere so far from equilibrium that for millions of years the planet hovered on the edge of sterility.

The conditions that had once made Earth exceptional nearly collapsed into the same silence that reigns across the rest of the cosmos. What matters most is that the rules have not changed. Earth today plays by the same laws it did 252 million years ago. We are not summoning lava fields the size of Europe, but we are altering the same variables, carbon in the air, oxygen in the oceans, the balance of climate systems that sustain us. The rise in greenhouse gases that once took hundreds of thousands of years, we have achieved in two centuries. The scale differs, but the mechanism remains the same.

It is not hard to imagine an Earth that never found its footing again, losing life altogether and extinction rates rising toward total loss, an Earth that once again joined the silent company of Mars and Venus. But it didn't. The Permian crisis passed, ecosystems rebuilt, and life's story continued. The world of the Permian before was no straightforward journey, though; this was still a world in which only the most enduring had a chance.

The World Before Destruction

The Permian summers seared, and the winters cut deep. Droughts stretched long enough to empty rivers. When storms came, they were devastating torrents that tore soil from slopes and swept it into valleys. Forests that endured were wiry and needle-leaved, their roots clawing at deep water. The air was drier; the climate severe. This was still a world of scarcity.

Amphibians lingered, but struggled more than before with fewer rivers and swamps. Their eggs and skins tied them to water, and in a drying world, that tether became a chain. Reptiles, and especially the synapsids, took their place.

The synapsids were not dinosaurs; those lay far in the future, but they were pioneers. Their skulls had a single opening behind the eye, a minor change in bone that enabled stronger jaws to close with greater force. From that advantage, an empire emerged.

They came in many forms. Early synapsid predators stalked the river margins, sails rising like blades into the morning light, warming their blood before the hunt. Herbivores grazed the low scrub in parallel forms. Therapsids followed, shorter-faced, more upright, with teeth that bit and tore, their broad, barrelled ribs driving a breath that foreshadowed the warm-blooded pulse of mammals. Some may have nursed their young, carrying the pattern of our ancestors in their bodies long before any true mammal walked the Earth. In the Permian, the line that would become us took its first steps.

Plants were reshaped by the shifting world. Conifers and other seed plants with tough cones spread into drier landscapes, and ginkgophytes and seed ferns persisted in wetter soils. In the seas, a rich cast of life endured. Shark-like fishes occupied different niches; early coiled cephalopods, such as ammonoids and nautiloids, moved through open water; and reefs built by sponges and algae offered shelter to fish. Even as climate patterns shifted toward greater seasonality on land, the oceans still shimmered with colour and motion.

The Permian stretched almost as long as the distance separating the last dinosaurs from humanity. From the outside, it looked steady, as if life had found a rhythm again. Synapsids ruled the land, forests survived in harder, drought-tolerant forms, and the seas were filled with predators and prey. It seemed an equilibrium that, while challenging, could last.

The surface calm hid pressure building far below. Magma pooled beneath Siberia. The crust of the Earth strained. No creature of the Permian could sense what was coming. They lived, hunted, bred, and thrived through their lives, constantly adapting to the environmental pressures.

But in the deep rock of Siberia, the fuse was already lit: an eruption that would wipe out nine in ten species and leave the Earth gasping, closer to lifelessness than at any other moment in its four-billion-year history.

21

The Bedrock of Industrial Might

Carboniferous (≈359–299 million years ago)

Three hundred and twenty million years ago, the tropics were drowned in green. A continuous belt of swamp forest stretched across the equator, its canopy so thick that daylight filtered down in a permanent dusk. Trunks stood shoulder to shoulder, rising like columns through a haze of spores and water vapour. Vines and fronds meshed overhead into a ceiling of damp foliage, dripping in slow, ceaseless rain. The air was stifling with the scent of wet plant matter and the hiss of falling condensation. Beneath that dark roof, every fallen branch and leaf sank into black water and soft peat, building the vast stores of carbon that would one day harden into coal.

All the while, Gondwana cooled in the south. Ice sheets multiplied, creeping across a supercontinent even as the tropics drowned in heat and shade. The planet held two extremes: a sweltering equator and a frozen pole. Beneath them, the swamps worked on burying carbon in seams of coal.

Those forests left a planet completely altered. Oxygen increased in the skies, creatures swelled into giants, fire stalked the land, ice conquered the south, and vertebrates carried seas inside their eggs. Its buried legacy still fuels our cities and still drives our engines to this day. North American geologists divided this era into the Mississippian and Pennsylvanian, yet elsewhere, it is known as the Carboniferous. For simplicity, I'll follow the broader usage.

A World Grown Dense

Wet forests and swamps covered much of the planet during the Carboniferous. The closest comparison today is the Amazon; even so, the likeness is only partial. Though the Amazon boasts a dazzling array of life, with tens of thousands of species vying for sunlight, its trees decompose and return to the earth within a year. Because little decayed in the Carboniferous forests, they were likely far denser, though less diverse.

Hollow trunks of giant lycopsids rose thirty metres before their crowns dissolved into mist. Their bark carried a lattice of diamond scars, each the trace of a shed frond. At the base, rootlets splayed outward like stilts, locking the trees into mud that never hardened. Horsetails thick as a man's thigh stood among them, their ribbed stalks built from endless, bamboo-like joints. Seed ferns pushed up woody stems and unfurled fronds wider than a person's reach, leaves beaded with constant condensation.

The atmosphere held nearly significantly more oxygen than today, and hung heavy with drifting spores; each breath tasted of wet cellulose and crushed greenery. Beneath the waterlogged floor, fallen stems and fronds formed a rising quilt of peat. Lignin, which evolved and remained beyond the enzymes of most Carboniferous fungi, sharply slowed decay amid the rain of organic matter. Year after year, the weight of swamp water and vegetation pressed those layers downward, the first seams of coal taking shape in the dark.

That locked carbon helped drive the oxygen surge that let dragonflies stretch to a child's arm span and fed fires fierce enough to flare even in a world this wet.

Over thousands of years, fallen vegetation became metres of peat. In millions, under floods of silt and sand, the peat was buried deep enough for heat and pressure to turn it into coal. Over tens of millions of years, the cycle repeated, seams stacked like pages in a book. Some stretch for hundreds of kilometres, the largest carbon vaults in Earth's history. As a result, the forests changed the air. With carbon pulled from circulation, oxygen built up in the sky to levels far higher than today. Fires raged more easily, kindled by lightning and fed by drier seasons.

The Carboniferous forests persisted for around sixty million years—just under the span of time between the last dinosaurs and us. Entire landscapes woven in green, burned, collapsed, and returned. Fossil evidence paints the picture in amazing detail, considering how old it is. Floors of petrified stumps frozen in place, roots still radiating into the soil; lenses of mineralised peat preserving leaves and seeds at the cellular level. The Carboniferous was the greenest chapter of deep time, a planetary engine that turned sunlight into biomass faster than decay could return it to the air.

When the Air Could Burn

Because of these forests, oxygen had surged to the highest sustained levels in Earth's history. It sits at around 21 per cent today, but the Carboniferous is estimated to have reached over 30, perhaps up to 35 per cent.

Under such conditions, widespread fire was inevitable. Above 25 per cent oxygen, and lightning strikes are considered kindling. Above 30, even sodden wood burns. Flames climbed the lycopsid trunks like matchsticks, crowns bursting one after another in cascades of sparks. The sound would have been violent—trunks cracking, the air roaring as it fed the blaze. Peat beneath the surface caught alight, glowing for months, while the canopy smouldered above.

These fires were woven into the balance of the Carboniferous. Flames recycled nutrients, opened ground for fresh shoots, and stopped the swamps from overrunning entire continents. Fire became a kind of ecological governor, setting limits on the very forests that had pushed oxygen high enough to burn them.

A fraction more oxygen, and firestorms might have scoured the land to ash. A fraction less, and the swamps could have marched on, burying even more carbon, dragging the planet deeper into ice. Excess oxygen made the Carboniferous a world of flames. Excess carbon dioxide shapes ours. In both cases, the message is blunt: change the air, and you change the world.

Atmospheres of Excess

In the charged air of the Carboniferous, the rules of size shifted. Insects, bound to the chemistry of their surroundings, grew into creatures of surprising magnitude. It entered their bodies through tracheal tubes that branched like rivers through a floodplain. With every ventilation, oxygen coursed deeper into their tissues, fuelling growth that defied their own previous design.

The confines of the atmosphere usually keep them small, but the changes now drove them to excess. Above the canopy moved *Meganeura*, a dragonfly-like insect with a wingspan of 65-70cm. They saw through compound eyes, bright with facets, to map through the shadows of giant ferns.

On the ground, another giant scurried between the trunks. *Arthropleura*, armour-plated, stretched to lengths greater than a tall man lying prone. Each step of its many legs sank into mats of rotting vegetation. Gigantism swept across the insect world like a viral infection. Scorpions 75 cm long stalked the margins of pools, their claws snapping and their stings poised.

Cockroaches grew broad-backed, their carapaces etched with ridges that gleamed in filtered light. Even silverfish and beetles, inflated beyond recognition—small creatures we know today, exaggerated into unfamiliar scales.

These giants leaned on a chemistry that could not last. When the atmosphere thinned, the insects receded. But on the margins of swamp and fire, other lives were testing a new solution. Amphibians crowded waterways, their eggs anchored to damp ground. Their reign on top was broad but tethered, each generation chained to ponds and pools. Beyond them stirred a lineage that would no longer bow to the reproductive restrictions in water.

Membranes of Independence

When the giants of the air and understory dwindled, the swamps were still alive with noise. Amphibians slid through the Carboniferous pools, their muscular bodies sheathed in a wet sheen, their broad heads armed with rows of teeth. Some were no larger than a man's hand, slipping unseen through reed beds. Others sprawled several metres long. They lurked in shaded pools, then burst upward in explosions of spray to seize whatever drifted close. The Carboniferous forests sang with their calls—deep croaks, rattling trills, the chorus of lives tethered to the wet.

They were abundant, diverse, and for a time unchallenged. Amphibians filled the ecological breadth of rivers, ponds, and wetlands. Some had limbs sturdy enough to drag them across damp ground between pools, but all were bound to water. Their eggs, jelly-soft and vulnerable, needed to develop and split in liquid. A drought could erase entire generations. However formidable they seemed here, their dynasty was restricted to the edge of the swamp. The constraint shaped them as much as it limited them. Their bodies bore the marks of compromise: lungs for air, but skins that absorbed water and lost it just as fast. When the surrounding forests smoked in lightning fires, they retreated to water, tied to a solvent that promised survival but also confinement.

Then came one of the most consequential innovations of the age—an embryo carried within its own private ocean. Inside a leathery shell, protected from desiccation and buffered from fluctuation, early life could unfold without the need for the consistent presence of water. The amniotic egg carried membranes that held fluid, stored food, and managed waste. It was a self-contained nursery, a capsule of independence.

With that single step, vertebrates could spread onto drier ground, following the mosaic of forest and clearing left by fire. They could withstand the droughts that closed ponds to amphibians. They could carry their young inland, into landscapes where water was scarce but opportunity was abundant.

The first reptiles were small; they slipped between tree roots, hunted among the ferns, and laid their eggs in places no amphibian could reach. Their young emerged formed from the eggs, miniature versions of their parents, ready to run and feed without the vulnerable stage of gilled larvae. This independence freed them not only from water but from the ecological dominance of the amphibians themselves.

From these beginnings, new lineages would rise—reptiles that would one day tower as dinosaurs, and others that would refine themselves into the ancestors of birds and mammals. The egg they carried was an escape strategy written into biology, a ticket to worlds denied to their competitors. As the climate cooled, ice spread across Gondwana, and swamps ruptured. As the stability of the wetlands faltered, amphibians suffered most. Their numbers fell, their diversity narrowed, and their reign slipped. The reptiles endured; their eggs weathered the drying of the land in the Permian. Their bodies carried them into spaces amphibians could not follow.

The amniotic egg was small, soft-shelled, and easily overlooked in the fossil record. Yet, the invention reshaped the history of life. Within its membranes ran a sealed system of water, oxygen, and waste exchange that freed reproduction from open water. In those first reptile eggs lay the survival strategy from which every bird, mammal, and reptile would one day descend.

The Great Fracture

The first cracks in the Carboniferous climate would have been subtle. Swamps that had once linked basin to basin left exposed patches of dry mud. Forests that had grown as a single fabric loosened into patches. From above, the once-continuous green would have looked like scattered islands in a widening sea of open ground.

The change came from the south. Gondwana slid across the pole, and glaciers swelled and shrank with the planet's tilt. With every advance, more water was locked into ice. Seas withdrew, the tropics dried, and the equatorial swamps that had fed the forests shrank away. It advanced steadily, but it did not yield. Each expansion crept deeper until the forests could no longer recover between the glacial intervals.

Coal seams, once laid in succession, are broken by barren strata. Peat gave way to sand and silt. Life inside the forests faltered as continuity broke, changing the air. With fewer forests to bury carbon, oxygen slipped from its heights, and fire lost its grip. Skies that had once burned with storms of flame steadied. The atmosphere that had once inflated insects and lit forests withdrew into a more balanced state. Today's tropical forests shrink in much the same way—by attrition, fragmentation, and isolation. Patches survive, but the ecosystem weakens; species vanish when corridors close. The Rainforest Collapse closed the age of swamp excess. It gave way to the Permian period's harsher and more divided conditions, where reptiles took root and amphibians lingered in decline.

The Coal in Our Engines

The Carboniferous energy deposits would resurface in the story, laying down black strata that stretch across the world. Heat and pressure transformed the peat into coal, and cell walls reshaped into a carbon lattice. It was sunlight in storage that humans would eventually learn to utilise.

At first, coal was just another fuel, stacked in hearths beside wood, its smoke curling from cottages and towns. But coal burned heavier, hotter, and longer. It could drive pistons, turn turbines, push ships across oceans and locomotives across continents. The Industrial Revolution began with the coal deposits feeding furnaces that turned iron into steel and boiled water into steam. Cities rose on it, railways threaded the map, and factories churned with production, fuelling the industrial age.

Burning it, they channelled another age's chemistry into the present. A forest that once pulled carbon from the air to swell oxygen high above the Earth now releases that same carbon back into the sky. What took millions of years to deposit and form is burning in centuries. Those forests created breathable air and stabilised the climate; in burning their remains, we begin to undo both. Even the plastics and chemicals that our lives depend on begin with carbon pulled from a swamp three hundred million years ago.

Intelligence arose, and with it curiosity, ingenuity, and hunger. If a species gained the ability to manipulate natural resources, it would inevitably exploit these seams. Coal was too potent a gift, too concentrated a store of power, to lie untouched forever. The Carboniferous planted the seeds of our climate-altering Anthropocene. Forests that once ruled from the poles to the tropics scripted the energy future of a primate that would come into being hundreds of millions of years later.

All over the world, the stored carbon of the Carboniferous fuels ships, trains, factories, and power grids. The forests that once spanned the equator now run through steel mills and toy factories, their ancient carbon released in a few human centuries. The abundance of this age built the modern world; its return as greenhouse gas now tests whether we can live with the consequences it leaves behind.

The seams fall open, the stored light escapes, and the gift of an ancient age becomes the trial of our own time. What began in the swamps of lycopsid trees hundreds of millions of years ago now tests whether a self-aware species can master the consequences of fuelling the fires that it rekindled. The energy of the forests remade the atmosphere once; in our hands, they may do so again.

22

Claiming Terra Incognito

Silurian–Devonian (≈443–359 million years ago)

The Silurian world was harsh and bare. Land stretched under a vast sky, its coasts made of raw stone with thin mats of moss clinging at the edges where water met rock. There were no forests on the horizon, no soil to hold the ground together, and no insects flying in the air. In the seas, reefs built by sponges and corals grew tall—types now extinct—and fishes began developing strong jaws for the first time. Water still covered most of the planet, but here life started its move from water to land.

By the Devonian, eighty million years later, the consequences of that leap would have been impossible to miss. Trees rooted themselves in the soil, spreading inland from coastal regions. Rivers that had once meandered without direction were held in place by roots. Fishes diversified into every shape: plated giants, swift hunters, and bottom-feeders. Out of them came the first vertebrates to heave themselves onto land, lungs opening where gills could no longer suffice. Insects took to the land, and the surface of Earth more closely resembled a world we would come to associate with life.

The Silurian and Devonian are presented as separate chapters in the geological calendar, but the story benefits from being told as one. During these epochs, plants and animals cracked the essential puzzles of terrestrial life: breathing air, resisting desiccation, and moving and reproducing away from water. Life figured out how to navigate the physics of living on land and to turn energy into movement, breaking free from the confines of water. It was the planet's first crossing into its own *terra incognita*—life claiming land no organism had yet mapped.

Pioneers of the Land

Four hundred and forty million years ago, the only colours of the continent were shades of grey and brown, broken by the white shine of wet rock. It would have felt more like the face of Mars if it had an ocean than the Earth we know. Life had already found success in these ancient oceans. But above the tide-line, nothing.

Then, photosynthesising organisms configured the expansion onto land. Small, moss-like plants clung to cracks in stone, their tissues absorbing moisture from rain and spray. Each patch held grit, slowed the wind, trapped a film of water long enough to establish presence. For millions of years, they crept outward, knitting damp rock into a patchwork of life. Their presence combined minerals with organic matter, creating the first soils. Dust that once drifted is now bound in place. The surface of continents transformed from inert crust into a medium capable of holding and recycling nutrients. The chemistry of the land leaned in favour of biology.

As the Silurian ended and the Devonian began, plants developed stronger stems that stood upright and waxy surfaces to retain water. They sent roots into cracks, breaking apart stone and reaching water below. They evolved xylem and phloem—internal tubes—to move water upward and sugars downward, giving plants height and strength.

Archaeopteris, a tree-like Late Devonian plant, grew ten metres high with broad, fern-like leaves. Its roots reached deep into floodplains, altering rivers and absorbing nutrients. Plants drew down carbon dioxide and released oxygen, transforming the land over millions of years. Continents once as lifeless as the Moon became coated in varying hues of green.

Living on land required solving problems that the ocean had always provided solutions for: buoyancy, hydration, temperature, and radiation. Plants engineered their way across these boundaries that were ripe for the taking. In learning how to operate outside of water, life bent into the very forms that could thrive without it. The green frontier is easy to overlook, but it represents one of the greatest reorientations in Earth's history. For the first time, the continents were part of the biosphere.

Before Any Bird Took Flight

As plants crept across damp rock, other small pioneers would follow, creatures whose bodies bore the first designs that could withstand the land's conditions. Insects were among the earliest to claim it, their jointed legs and sealed skins allowing them to cross ground where vertebrates had not yet set foot.

Insects had skeletons on the outside, made of chitin, a tough, flexible material, that kept their bodies moist and safe from the air. These shells had segments, hinges, and spines. Inside, a system of tubes called tracheae branched through the body to bring in oxygen, so insects did not need lungs. While vertebrates struggled at the edges of land, insects already had what they needed to survive dry air, strong sunlight, and changing winds. They fed on decaying plants above the shore.

Later, some insects began to eat living plants. During the Silurian and

Devonian, the earliest insects had no wings but already had strong jaws for chewing and mouthparts for scraping spores. They also had claws and simple guts for climbing and eating tough plant tissues, long before some later insects developed specialised mouthparts for sucking. These adaptations began the long relationship between plants and insects. Plants evolved defences against being eaten; insects adapted in turn, creating a feedback cycle that would shape terrestrial ecosystems over deep time.

Then came the most radical step of them all: flight. Long before any bird, long before any bat, wings would unfold from insect bodies, delicate membranes stretched between veins of chitin, their muscles tethered inside the thorax like pulleys. By today's evidence, insects did not fully master powered flight until the Carboniferous, but the anatomical groundwork for life in the air was laid far earlier, as bodies already adapted to dryness, exposure, and open space found new ways to move through it. With wings, insects could travel between habitats, escape predators, and pursue mates. They colonised the sky; no vertebrate would follow them there for another hundred million years.

Small bodies, rapid reproduction, and flexible diets gave them even more resilience. They could adapt faster than the slower, bulkier vertebrates, still finding their way. Insects endured environmental swings that erased other lineages, persisting through upheavals that reshaped life on land and sea. They followed into every niche plants could offer.

Even now, hundreds of millions of years later, our world is only possible because of their ascent. Insects pollinate crops, recycle nutrients in forests, and transfer energy through food webs. Their numbers vastly exceed ours. They first mastered land, and in many ways, never surrendered it.

The First Footprints in Mud

Before the Silurian, most vertebrates fed by grazing on surfaces or filtering tiny organisms through soft mouths. But in the ancestors of jawed fish, the cartilage that supported the gills began moving forward. By the late Silurian, these supports had turned into real jaws—muscles pulled, plates met, and the water gave way. Bones at the edges of jaws hardened into teeth, some with sharp edges. The seas, calm for ages, now had true predators. Fish could grab prey rather than just waiting. Mouths became tools for crushing shells or slicing flesh. This sped up the flow of energy through food webs.

By the Devonian, this change was much further along. The seas teemed with bizarre predators and prey. Placoderms swam in armour-plated bodies, filling estuaries and bays. *Dunkleosteus* hunted among them with jaws able to close like a guillotine, cutting through bone and muscle in an instant. Reefs became crowded fortresses made of stone and skeleton, standing against attacks.

During the late Devonian, a body lay half-submerged on pebbled sediment. Its head was broad, the eyes perched high on the skull, staring outward from both water and air. The body was still fishlike. Inside the fins, bones were arranged in a pattern we now recognise as shoulder, elbow, and wrist. This extension of the body did not move with the sway of a fin. It moved with the leverage of a limb.

The fossil we know as *Tiktaalik* carried both gills and primitive lungs. It drew oxygen from the water, yet could also raise its head and gulp air. Its ribs were stronger than those of its ancestors. They held their chest clear of the mud. The skeleton shows a creature capable of hauling itself onto banks in shallow water, where dissolved oxygen was too low for most fish to thrive. Its movement would not have been graceful. It could not have roamed far. Even these clumsy efforts marked the first evidence of a vertebrate body supporting itself, however briefly, on land.

The Devonian world offered a reason to attempt such a crossing. Shallows dried into pools during seasonal droughts. Oxygen dropped in stagnant channels. Food lay stranded in muddy margins, where insects, worms, and other invertebrates had already established themselves. A fish that could prop itself on a bank to breathe and snap at prey had an advantage. In that advantage lay the beginning of a new kingdom. Every tundra, mountain, and desert ever walked by vertebrates can trace itself back to these first awkward excursions from the shallows.

Reeftropolis Lost

In the Silurian and Devonian seas, life built the first ecosystems with vertical and horizontal dimensions. Corals and stromatoporoids layered their skeletons into ridges, ledges, and walls. They transformed flat seabeds into reef systems that sprawled for hundreds of kilometres. Light fractured through crevices. Tides slowed as they met these barriers. The seabed rose as a metropolis wrought from coral and ancient bone.

Each reef was a composite of life turned to stone. Generations of corals and sponges cemented themselves atop one another, their skeletons fusing into limestone buttresses. The dead became the foundation for the living, a cycle repeated for tens of millions of years. Over time, the reefs altered local chemistry, redirected tides, and amplified the seas' diversity.

Within their alcoves and channels, trilobites searched for food, brachiopods clamped to rock, and fishes threaded through shaded corridors. Encounters between predator and prey became unavoidable, compressed into spaces where hiding was difficult. In this density, evolution accelerated. Responding to threats and opportunities that lurked behind each outcrop.

By the height of the Devonian, reefs stretched across continents. They stood as submerged fortresses, layered with corals, molluscs, crinoids, trilobites, and shoals of fish. The fossil record preserves their richness with startling clarity. Limestone beds in places like Western Australia reveal reefs built over millions of years. Some preserve entire communities in cross-section: corals branching upward, stromatoporoids layered in tiers, shells scattered across sediment, fish teeth fossilised in situ.

Ocean conditions were altered in the late Devonian. The climate cooled and then warmed again. Sea levels rose and fell, fragmenting habitats. Oxygen drained from the depths as circulation slowed. Algal blooms surged in shallow waters, feeding microbes that stripped oxygen from the sea when they died. Collapse came in surges rather than in one instant. Extinction events rolled through marine ecosystems across twenty million years. Reefs were hit hardest; whole provinces dissolved into rubble that did not rebuild.

Corals would return, but never again at the same scale or with the same dominance. They had amplified the evolutionary tempo of the seas, but their very complexity became their weakness. When the oceans slipped into oxygen crises, the density of life that had thrived in confined quarters became unsustainable.

The Silurian and Devonian reconfigured the planet. They clothed bare continents in vegetation, and armed vertebrates with jaws that would ripple through every future food chain. They built reefs that rivalled continents in scale, then watched them collapse under the strain of oxygen-starved seas. From these upheavals came the template of life on land and the structure of ecosystems as we know them. Beneath these layers lies an older foundation, where trilobites flourished, and the first great reefs took shape, only to be tested by the earliest of Earth's mass extinctions.

23

Webs Beneath the Waves

Ordovician (≈485–443 million years ago)

Long before forests scaled the continents or insects swarmed the skies, the ocean held the world. The land was rock, shifting only through the slow machinery of geology. Meanwhile, seas spread across vast expanses—warm and high—flooding deep into the interiors of half-formed continents.

The Ordovician was an age when water dictated almost everything. Liquid water dissolves nutrients, transports waste, and carries larvae to new ground. Buoyancy lifts bodies against gravity. Currents deliver food and maintain genetic exchange across distance. The sea is a medium in which life could experiment.

Currents of Creation

In the Ordovician, the sea became the most elaborate laboratory the planet had yet known. Gradients of light, oxygen, and temperature carved thousands of ecological chambers. Currents linked these, carrying larvae and drifting life from one to another. Delicate bodies survived because water held them. Larger bodies grew because buoyancy relieved them of gravity's pull. Currents acted as circulatory systems. They carried plankton blooms across vast reaches, sweeping energy from coastal shelves to open waters. Along rising mountain belts, sediment poured in from eroding highlands. Periodic influxes of phosphorus and calcium bolstered food chains from the bottom up. Storms stirred the surface, mixing nutrients into light and forcing organisms to adapt to disruption. Nothing was still; every particle moved.

Existing in this medium meant being permeable, prone to leaks, and reliant on movement. Each body plan reflected a negotiation with water: how to float, how to filter, how to resist drag, how to seize what drifted past. Physics governed form further. Creatures could grow long or broad without snapping under their own weight. Many shells followed a spiral growth pattern, maintaining strength as animals enlarged. Appendages and body profiles shifted toward forms that moved and fed more efficiently in water. Eyes, where present, adapted to the constraints of light underwater—sensitivity, field of view, the dimming of depth.

This obedience to physics produced stability of a kind. For the first time, ecosystems reached a density where roles multiplied. Grazers cleared mats so that filter feeders could thrive. Filter feeders altered currents, shaping what reached neighbours downstream. Predators struck at prey that, in turn, hardened their shells. Every success reshaped the conditions for another.

Understanding the importance of those initial oceans requires acknowledging how rare they are on Earth compared to other planets and how long they lasted. Mars' atmosphere thinned, pressure dropped, and what water remained either froze or drifted into space. Closer to the Sun, Venus suffered the opposite fate; its seas boiled. Water vapour built up in the air, trapping still more heat until evaporation fed on itself. The icy moons of Jupiter and Saturn may still hold liquid seas beneath kilometres of frozen crust, but sealed away from the sky. As far as we know, only Earth carried oceans vast and open, heated by sunlight from above and stirred by tides below.

The seas revealed that complexity itself can become self-sustaining. When networks of interaction reach sufficient density, they feed their own expansion. Every lineage that would later walk the land or fill the air first depended on structures established here, in water.

The Great Ordovician Biodiversification Event

For three billion years, webs of life had remained modest—microbial mats rippling across tidal flats, simple shells scattered on the seabed, algae drifting in suspension. The Cambrian period, which we will explore in the next chapter, opened an ecological door, and the earliest armoured predators emerged. Yet these were still loose assemblages, easily undone. In the Ordovician, oceans crossed a gradual point at which life ceased to be a handful of experiments and became a dense, entangled fabric. This surge, the Great Ordovician Biodiversification Event, or the GOBE, was a long ignition, a rewiring of the biosphere into networks of elevated density.

Across continental shelves and flooded interiors, species multiplied until ecosystems knotted into intricate webs. Trilobites, once uniform, split into burrowers, swimmers, scavengers and predators. Brachiopods, marine invertebrates resembling clams, hardened their shells. They crowded rocky ledges in ridges of calcite. Molluscs pushed into new lineages. Some coiled into cephalopods with shells that stretched metres long. Their tentacles steered them near the top of their food chains. Bryozoans spread like lace across surfaces. Colonies of living mesh altered currents and trapped sediment. Crinoids rooted themselves on the seafloor and unfurled feathered arms to sieve the passing tide. Beyond these communities, reefs rose in places. Varying mixtures of sponges, bryozoans, algae and some early corals built them, layering skeletons into walls, ledges and buttresses. Each new structure created niches. Each niche summoned new life to fill it.

The catalyst for this transformation was geology. Sea levels stood high, flooding continental interiors with shallow, sunlit waters that stretched for thousands of kilometres. Over time, changing climates and circulation sharpened gradients between warm surface layers and colder depths. These changes created differences in temperature, oxygen, and salinity. Evolution thrives on gradients. Every gradient difference opens a new ecological space.

Once ecological networks reached a certain density, they reinforced themselves. Scientists describe this growth in the language of systems theory: emergent complexity, feedback loops, self-organising networks. In practice, each success created conditions for others. Filter feeders stirred currents that nourished scavengers downstream. Burrowers mixed sediments that fed algae above. Predators forced prey into new strategies. These strategies, in turn, created new predators.

Before the GOBE, life had been shaped primarily by the environment—chemistry, geology, and climate. Afterwards, ecosystems themselves began to shape the trajectory of evolution. Waste from one lineage became fuel for another. Predation sculpted armour. Armour-sculpted predators. Entire lineages bent in response to one another's presence. The Cambrian explosion filled the seas with trilobites, worms, sponges, and the first predators with hardened skeletons. But those ecosystems were loose, fragile, and scattered across shallow seas. Their reefs were small, their niches broad, and their webs thin.

From this point forward, the web itself increasingly dictated what could survive. Ecosystems often reassemble into familiar roles and structures. Certain configurations of life are more stable than others. As the weave grew more intricate, its vulnerability to destruction intensified. The Ordovician seas would learn this lesson in the coming mass extinction.

Complexity's First Collapse

By the Late Ordovician, the seas were crowded, their networks taut with exchange. Around 443 million years ago, Earth lurched into a cold. Ice spread across Gondwana, and glaciers crystallised. Sea levels fell by more than a hundred metres. Warm inland seas that had sheltered life drained to nothing. Shelves once crowded with reefs became exposed rock, where entire ecosystems had stood. The water's chemistry shifted in parallel. Evidence suggests oxygen retreated from the depths until many regions likely became stagnant. Zones that once brimmed with movement fell still.

The first wave of glaciation erased habitats wholesale. When the ice retreated, flooding returned. This destabilised the seas again, stripping nutrients and drowning coastlines. Organisms adapted to narrow ranges of depth or temperature were the first to vanish. Predators lost prey, filter feeders lost currents, and reefs crumbled. The cascade spread outward from every vacancy.

By the end, about 85% of marine species were gone.

Trilobites, once the ubiquitous sculptors of the seafloor, were devastated. Entire families of forms disappeared. Graptolites, drifting plankton that had dominated the open seas, dwindled to a fraction of their former diversity. Reef systems collapsed. It took millions of years to recover. Life's first grand web was torn through the very connections that had made it resilient.

This extinction would echo through every other crisis to come. Complexity both stabilises ecosystems and increases their sensitivity. Diversity weaves resilience, but the same interdependence makes the system vulnerable when foundations shift. The tighter the weave of a web, the more one rupture can unravel it. In this sense, the Ordovician mass extinction was unique. It was the first time life was undone by the very success of its proliferation. Evolution laced the seas with bonds. Those bonds then undermined the stability of the biosphere.

After the extinction event, ecosystems bent back into shape. New players filled old roles. Patterns of interaction re-emerged because natural selection reinforced structures most able to persevere. Recovery was not swift; it took millions of years. Still, the underlying interplay of ecosystems weathered the storm.

Fossil beds of this time preserve abrupt ecological turnover. Reefal communities thinned. Once-common fossils dropped sharply in abundance. The strata above tell of recovery—new lineages radiating into emptied spaces, ecosystems knitting themselves back into coherence. Evolution moved in loops. It stumbled and resumed, each collapse carrying the seeds of a different beginning.

24

The Bakuhatsu Upheaval

Cambrian (≈541–485 million years ago)

The Ordovician waters were a tight lattice of species. Beneath that layer, the structure slackens; the weave of life lies looser. Here, the Cambrian Explosion erupts. Before, in the Ediacaran, life lay pressed against the sand like fabric, drifting, ingesting without mouths, absorbing without eyes. Now, the surface of the seabed is cross-hatched with trails. Creatures scuttle, twist, burrow. Their signatures are preserved in the sediment.

The water is shallow and warm. Across the seafloor, a group of trilobites advances, legs shuffling in synchrony beneath. They pause at the mat, limbs combing the silt for food. Above them, the light darkens. A shape passes, and the throng breaks. Some roll into tight curls, shells locking shut. Others scatter sideways into grooves they have carved before. The shadow slides on, and the swarm re-forms, as if nothing had happened. Evolution had built bodies that could go beyond growing and replicating; now they could sense and react. Predators were new, armour provisional, senses freshly minted. The ocean became an arena of trials, mistakes, and sudden success.

In the layers beneath the Cambrian, fossils are faint and uncertain, soft outlines of quilted bodies, almost erased. Step upward into Cambrian beds, and life had hardened. Harder materials leave traces we find today. This turning point is often called the 'Cambrian Explosion', but that metaphor of a sudden blast is misleading. Rather than an instantaneous burst, the Cambrian was a prolonged ignition—a slow *bakuhatsu* (the Japanese word for 'eruption'), an upheaval of invention that remade the seas. Over 20 million years, armour, eyes, muscles, and jaws assembled gradually. With the onset of predation and defence, passive existence ceased. This is the world we enter: an ocean environment where organisms tried new strategies, most failing, but some persisted for 500 million years through their successful descendants.

Explosions in Slow Motion

The Cambrian explosion resembled a fire slowly advancing through dry grass—creeping, looping, and igniting patch after patch until the entire plain was ablaze. This comparison emphasises the gradual rather than instantaneous nature of the event. Twenty million years, though a long time, appears as a brief flare in Earth's vast history. The main difference lay not only in the event's pace but also in the nature of the innovations that emerged.

One force was oxygen; for billions of years, cyanobacteria had been spilling it into the oceans, molecule by molecule, a slow and patient trickle. In earlier ages, it was a trace element, reactive and fleeting. But by the dawn of the Cambrian, it had reached levels that enabled new metabolisms to emerge. With more oxygen in the water, muscles could fire faster, and tissues could grow larger. A microdictyon scuttling over the seabed was a chemical event, wrapped in new biology and powered by a gas absent in older seas.

Another force was buried in the genome. Across many lineages, a new system of genetic instructions emerged—the Hox genes. They were the switches that directed where tissues should grow, when limbs should appear, what should become the head and what should become the tail. Before this innovation, bodies had been more like drifting fabrics, assembled without deep positional control. With Hox genes, a mutation in a single gene could shift the pattern of an entire organism. A limb might split in two, or an eye might appear. Oxygen provided the fuel, but the real breakthrough lay in the genetic code itself—the developmental programs that made new body plans possible.

Predation elevated the stakes. Once a hunter could chase, no creature could remain idle. The moment a predator closed in, the world was divided into survivors and failures. Prey scrambled into burrows, sealed themselves in armour, or darted into the water column. Each defence provoked a counter; speed met speed, claws met shells, spines met teeth.

The seabed itself joined the contest. Burrowers ripped open what was previously coated with calm microbial mats. Worms and arthropods churned the sediment, leaving behind galleries and spirals that aerated the mud. This changed the chemistry of the floor, drawing oxygen deeper, releasing new nutrients, and destabilising old habitats. The stage of life's drama became an active participant, shaping what could live above it.

More oxygen allowed muscles to pull harder. Hox genes rewrote body plans to make those muscles useful. Predation applied pressure to move, hide, or strike back. Burrows changed the substrate, creating arenas for new behaviours. The result was a self-reinforcing storm of invention.

The fossil record populates sharply after the Cambrian Explosion because there was more to preserve. With hard parts—shells, spines, skeletons—organisms left remains that could persist long enough to enter stone. The veil that had hidden so much of earlier history lifted. Over twenty million years, the oceans filled with new anatomies, a sudden flowering of forms that set the blueprint for nearly every animal body that lives today.

Hard Bodies

Biomineralisation, the biological craft of turning seawater's minerals into armour and framework, may have started as a regulation, an accidental by-product of cells managing excess ions. But what began as waste control, evolution seized and repurposed. In the contest of predator and prey, such adjustments were decisive. A soft worm could be pierced or swallowed. A worm lined with spicules could scrape free, or at least survive long enough to breed.

The ocean itself was ready to be harvested. Rain and rivers dragged minerals into the sea, saturating it with ions that had once been locked in stone. Calcium in particular swelled in abundance. With higher oxygen levels than before, animals could afford the metabolic cost of hard parts. Energy was available, raw material plentiful, and natural selection made use of both.

Some organisms secreted simple plates; others built elaborate frameworks. Tiny tubes lined the seabed like pipes. Cone-shaped shells bristled from the silt. Armour fused into broad shields. A lineage might try one material, abandon it, then return generations later with a unique mix. Trilobites embody this better than any other group. Their jointed exoskeletons were divided into ribs that flexed, plates that locked, and spines that jutted. They wore their environment as armour. And with every growth, they split their shells along a seam and stepped out, soft for a moment, vulnerable until the new armour hardened. The seafloor grew with these castings, layered into stone.

But hard bodies were not guarantees of success. The fossil record is littered with creatures who tried too much or too little. Some managed only thin cuticles, as fragile as eggshells. Others encrusted themselves in shells too heavy to move far, easy prey for whatever had the patience to pry them apart.

The larger story, though, was permanence. Once life drew minerals into its flesh, it never abandoned the habit. Hard bodies became the scaffolds of ecosystems. Predation gained a new theatre: crushing and boring, piercing. And with each adjustment, the archive deepened. For the first time, Earth's history became visible in bodies built to last.

The Beholder of the Eye

For most of Earth's history, life had found its way blindly. Cells sensed gradients of light or chemical traces, but no creature had yet assembled an image of the world. The seas were full of contact and reaction, but not of sight. During the Cambrian, evolution shaped a one-way threshold. Light-sensitive cells sank into cupped hollows, hollows tightened to pinholes, pinholes grew lenses able to bend light. The first true eyes emerged from these gradual steps.

The invention was just a cluster of cells at first, able to distinguish between brightness and darkness, but even this was a significant advantage. A flatworm with a patch of light-sensitive skin could steer itself toward shade and avoid being cooked in the shallows. Over time, the patches curved inward, granting direction. Each adjustment magnified the advantage until animals could do more than sense light; they could build a picture that mapped reality.

Trilobites carried the most interesting version of these early experiments. Their compound eyes were crystal towers, built of calcite, the same mineral that makes limestone. Thousands of tiny lenses studded each eye, each one a prism bending light. The trilobite did not see a seamless view like ours. It saw basic evidence of movement, differentiation of light and shadow.

The impact of vision cannot be overstated; to see was to expect. A predator could now track prey before striking. From vision flowed a cascade of behaviours. Eyes allowed courtship displays—flashes of pigment to attract, dances to impress. They allowed mimicry—one animal borrowing the form of another to deceive. They allowed recognition, memory, and navigation. The world became patterned with signals.

The eye also opened the fossil record itself. In Burgess Shale slabs, the dark stains of compound eyes still glisten faintly, circles packed with facets. Some of the most detailed fossils we possess are preserved because the eyes were mineralised. Trilobites, in trying to see more clearly, left clearer evidence of themselves for us.

Every line you read, every constellation charted, every tool shaped by hand was guided by vision. Firelight was tended because eyes saw flame. Language was written because the eyes could follow symbols. Technology unfolded along the channel that light opened. The same cascade of photons that struck a trilobite lens now strikes your retina.

What began as a patch of pigment on a worm's skin became one of the most powerful tools evolution ever devised. It transformed the biosphere from a place of contact into a place of perception. The eyes of that age are gone, but the act of vision continues, unbroken, from their crystals to ours.

A 541-Million-Year-Old Controversy

The Cambrian takes its name from Cambria—the Roman word for Wales—where the pioneering geologist Adam Sedgwick mapped the first beds of this strange age. Sedgwick would later reject the theory of evolution outright, proposed by his student Charles Darwin. When Darwin published *On the Origin of Species* in 1859, the Cambrian loomed as one of the most awkward tests of his theory.

By then, the Cambrian had already been drawn as a hard boundary in the fossil record. Below, though, the rocks seemed silent. Above this, they were crowded with trilobites, brachiopods, sponges, worms with jaws—a sudden festival of complexity. To Darwin's critics, the message was obvious; here was a line in stone where life began anew, not through gradual change but through divine invention.

Darwin admitted the problem with accountability. 'To the question of why we do not find rich fossiliferous deposits belonging to these assumed earliest periods prior to the Cambrian', he wrote, 'I can give no satisfactory answer'. He called it a 'serious difficulty'. For opponents, it was exactly what they needed: a hole in the argument. If natural selection worked only through gradual accumulation, why did the rocks not record the steps? Why did complexity surface abruptly, as if summoned?

Darwin's response was as daring as it was simple. He argued the fault was not in the history of life but in the remains in the stone. The ancestors were there, but too delicate to preserve. Life had been shaping itself long before, but until shells and skeletons evolved, it slipped away unrecorded.

The Cambrian strata became a battlefield over how the world itself should be read. Nineteenth-century Europe was still shaped by theology, by the sense that creation was ordered by design and overseen by God. To many, the abrupt Cambrian explosion matched that vision as evidence of the divine. Animals were brought forth all at once, without messy origin stories or ancestral fumbling. Darwin's gradualism, by contrast, unsettled the idea that nature followed a moral order.

For critics, its suddenness confirmed that evolution was a fantasy. For supporters, it posed a riddle worth solving. Darwin himself never saw the solution in his lifetime. The silence beneath the Cambrian followed him to the grave. For decades after, creationists held the gap as proof of Genesis. Museums displayed trilobites as if they had stepped whole into being. But by the early 21st century, Darwin's intuition had been vindicated.

What survives in the fossil record is determined by conditions that allowed the fossils to be preserved. Our knowledge is stitched from the remains we are lucky enough to find, and those fragments can mislead as easily as they reveal. Today, when we prise open a slab of Ediacaran sandstone and trace the outline of a frond or the swirl of a burrow, we see the evidence that would have reassured Darwin. They are faint, almost erased, but they show continuity where once there seemed a rupture.

For Darwin, the silence beneath the Cambrian was a torment. The rocks misled him for a time, but they also sharpened the question that drove him: how can life change and still be life? The Cambrian's abruptness forced that question into focus. Its resolution, uncovered in the soft impressions of the Ediacaran, invites us further down, into a simpler world.

25

Thaw and Surge

Ediacaran (c. 635–541 million years ago)

Around 635 million years ago, the planet thawed from one of its deepest freezes. Glaciers had reached the equator. Seas may have lain sealed beneath a frozen crust. When the ice finally eased, the oceans opened once more to the sun. Rivers carved raw continents and washed nutrients into the sea. Oxygen seeped through water that had long been sealed. With this release, life scaled itself into visibility.

For billions of years, existence had remained microbial, spread across the planet in mats, films, and colonies. Microbes transformed the chemistry of the atmosphere, pulled gases from rock, and inscribed themselves in stromatolites, but they remained too small to build a world that could be seen. In the Ediacaran, that changed. For the first time, the oceans carried organisms large enough to be visible. The earliest larger bodies appear to have maximised surface area, though their internal complexity remains debated.

What survives of them are impressions pressed into sandstone and siltstone. They are shapes without depth—almost like diagrams of bodies rather than structured anatomies. Some resemble ribbed ovals the length of a person's arm. Others are tethered fronds, as delicate as fabric but fixed to the mud. Still others branch like fractals, repeating themselves at smaller and smaller scales. They resemble nothing alive today; most belong to no known lineage that continued.

Our knowledge of these life forms comes through four windows scattered across the continents. None is complete, but together they reveal traces of a biosphere that endured almost ninety million years.

At Mistaken Point in Newfoundland, the volcanic ash of an ancient eruption smothered entire communities in an instant. This preserved the sea floor. Rangeomorphs stretch outward across bedding planes, their branches repeated. Quilted ovals overlap like tiles.

In the White Sea of northern Russia, the record swells with diversity. Here we see forms more varied—fans, quilts, segmented outlines, and for the first time, traces of movement. *Kimberella* appears in microbial mats. With it, run parallel scratches in the rock, which are considered feeding marks, deliberate and directional. They are evidence of an animal shaping its surroundings; a faint but unmistakable sign of behaviour.

Further south, the Nama Group of Namibia records the twilight of this world. Fossils here are fewer, their outlines thinner, and their communities sparser. Some slabs suggest stress—bodies disturbed, and assemblages fragmented. The record hints at a system stretched to breaking point.

In the Flinders Ranges of South Australia, another part of the story lies preserved. In the 1940s, while surveying for minerals, the young geologist Reg Sprigg stumbled across strange imprints in sandstone. He recognised their antiquity, but few others took notice. Only later were they accepted as evidence of life older than the Cambrian. Among these slabs lies *Spriggina*, a small segmented form with bilateral symmetry. Its structure gestures toward the future, echoing designs that would later thrive. Yet like

Kimberella, it left no confirmed descendants.

Taken together, these sites—Avalon, White Sea, Nama, Flinders—are the only reason the Ediacaran is visible to us at all. Without them, ninety million years of history would be erased altogether. What we see on these slabs is not much. Still, it's enough to reveal an entire biosphere—one that flourished, then vanished. The fossils are vivid. Some are so sharp you can run your fingers along their ridges. But the bodies they record grew in scale without gaining durability. They grew large without building inward.

Geometry of Survival

The pattern of Ediacaran life was pushed towards maximising surface area for interactions with the ocean water. They lacked skeletons to brace against gravity, guts to digest food, and nerves to sense danger.

Rangeomorphs increased their feeding surface through repeated folds and branches, drawing dissolved nutrients directly from seawater into their tissues. Geometry carried the logic of feeding, effective under stable currents.

Charnia rose as a tethered frond anchored to the sediment. Its tissues stretched outward like a banner into the current. It resembled a leaf but belonged to no plant. Its form was a sheet of living tissue, unreinforced by skeleton or vascular system. Dickinsonia grew by adding segments at its margins. Its design was all exterior, absorbing nutrients from the mats beneath it. In 2018, geochemists teased evidence of cholesterol molecules from a Dickinsonia fossil. This tilted the debate towards an animal identity. But its body plan lies outside any modern group. It remains singular, recognisable only as itself.

Such designs flourished because they suited their environment. Microbial mats offered firm support and nourishment, holding sediment together in thick layers. Organisms spread out over these mats, competing by growing. A frond that grew taller blocked the current first. A quilt that spread farther cast a shade on its neighbours. These changes showed that the organisms existed, but did not involve movement. They took place over thousands of years. In such long-lasting stability, unmoving forms could thrive.

But the organisms of the Ediacaran depended on this stillness. Their survival rested on the mats remaining intact, oxygen remaining steady, and currents remaining predictable. When burrowers churned the sediment in the Cambrian, the mats disintegrated. Flat passive surfaces offered no defence against probing, tearing predators. When oxygen levels fluctuated, diffusion through thin sheets was insufficient to sustain metabolism.

Even in this world of surfaces, hints of behaviour appear. Kimberella's feeding marks show agency, a body choosing its path and inscribing it in sediment. Later in the period, Treptichnus appears; these traces are widely attributed to worm-like bilaterians, animals capable of probing, reversing, and trying again. These marks record nervous systems coordinating behaviour, making decisions, and altering paths.

The paradox of the Ediacaran is that the clearest fossils are those whose legacies failed. Their surfaces were imprinted in stone, while the ancestors of future lineages left almost nothing recognisable. Chordates, arthropods, molluscs—all are inferred from molecular data to have existed in some form, though direct fossil evidence from this period is lacking. They may have been present, hidden among these bodies, too small or soft to preserve. The lineages critical to our story are absent; those that ended are vivid.

The Final Garden of Sand

The close of the Ediacaran is abrupt. In one stratum of rock lie fronds and quilts. A few metres above appear burrows, shells, spines, and eyes. Several pressures converged here. Oxygen levels, which had supported surface-dwelling organisms, fluctuated. The Shuram anomaly, the greatest carbon isotope disturbance in Earth's history, is interpreted as a signal of crippling instability in global cycles. Predators with directed movement appeared, probing and grazing. Each of these forces alone might have stressed the biosphere; together, they erased it.

The animals known from the Ediacaran period appear not to have adapted to the new conditions. Their surfaces and sheets offered little resistance to change, for much had not changed in their reign. They ended, and complexity was rebuilt along different lines—skeletons, muscles, nerves, eyes. The organisms of the Cambrian explosion that followed replaced them, though the exact causes of this transition remain debated. The stillness of quilts and fans gave way to the restlessness of predators and prey.

The Ediacaran holds organisms with no clear descendants and lineages with no clear ancestors. Though their forms are distinct, the fossils lack connection, appearing sharp yet incomplete. We can place a hand on sandstone in the Flinders Ranges and feel the ridges of a body half a billion years old, yet we cannot point to any as an ancestor of vertebrates.

Mistaken Point, White Sea, Nama, and Flinders are four sites that provide a window into a long-gone biosphere. Future discoveries could add new pieces to the puzzle. We might find more evidence, though a complete account will likely elude us.

What other lineages lingered unseen in those seas, essential to our story yet invisible now? We can only speculate. Fossils show flourishing life and its subsequent extinction. Its paradox is that we know it but cannot follow it forward. The Ediacaran is the ultimate instance when soft bodies convey messages to us in stone. Step deeper, and the record dissolves into isotopes and chemistry, into oceans sealed beneath ice. The stillness of quilts and fronds is only a prelude. What follows is the greater stillness of a planet locked in ice.

26

Fimbulvetr Aeon

Cryogenian (≈720–635 million years ago)

The sun hung weak above a white horizon, its warmth scattered, its rays dulled to nothing. Below, ice reached in every direction. Not like the glaciers we know today, broken and blue with meltwater, but sheets that spanned continents, kilometres high, pressing down with unimaginable weight. The only sounds came from the crack of ice and the howling of prevailing winds.

You would be forgiven for thinking this was Antarctica, but this was closer to the equator. Where Adelaide now bakes in the summer heat beneath eucalyptus trees, glaciers once capped the surface. The Cryogenian Period, beginning around 720 million years ago, plunged the Earth into its most extreme glaciations. Twice, the planet froze almost to its limits. The first, the Sturtian glaciation, endured for more than fifty million years. The second, the Marinoan, stretched for at least fifteen million more. Between them, the world may have thawed only briefly before sliding back into ice. *Fimbulvetr*—in Old Norse, the 'great winter' that heralds the world's end—and aeon, an age beyond measure, together evoke the time when Earth entered this long, deep freeze.

Some geologists argue the planet was never a perfect snowball but more like a 'slush ball'. In this version, ice still reached the tropics. Thin crusts or open leads of water lingered near the equator. These cracks would have admitted light and sustained photosynthetic life more easily than a sealed ocean. The distinction matters less than the scale. Whether snowball or slush ball, Earth had tipped into a state where survival depended on narrow corridors. The difference between a few open ribbons of sea and a global shell of ice marked the line between persistence and extinction.

In Namibia, we find evidence of this: solitary boulders sit stranded in fine-grained rock—dropstones carried by icebergs into seas and dropped as they melted. In Canada, pavements of bedrock lie scoured smooth, carved by glaciers that spread across what was once a shallow shelf. In South Australia, layers of glacial till lie among rocks that once formed in the tropics. Together, they tell us that Earth froze to its equator, oceans sealed beneath white lids, continents and seas hidden.

The world above the ice was void of life, and even photosynthesis may have faltered. The hydrological cycle stalled. Clouds dispersed, winds slackened, and rain ceased. Below the crust, in oceans held liquid by pressure, life held on, barely. It was microbial, but its resilience across tens of millions of years would prove essential.

A System in Collapse

A planet does not slip into a deep freeze by chance. The Cryogenian was the product of Earth's machinery tipping out of balance, with feedback loops that reinforced themselves until the system locked.

One likely trigger was the slow unravelling of Rodinia, a supercontinent like Pangaea, but far older. As it split, fresh rock spread across continental margins. Rain, slightly acidic with dissolved carbon dioxide, reacted with these surfaces and drew greenhouse gases out of the atmosphere. The carbon ended up buried in sediments; over millions of years, the insulating blanket around the Earth thinned.

As carbon dioxide dropped, ice formed; glaciers expanded across entire continents. White surfaces reflected sunlight back into space, cooling the planet even further. The albedo effect—where ice makes its own spread inevitable—gathered momentum. After ice extended to mid-latitudes, the planet triggered an inescapable feedback loop.

Other forces may have deepened the effect. Volcanic eruptions could have filled the sky with sulphur haze, scattering sunlight. Changes in the Earth's orbit might have reduced insolation at critical latitudes. Methane, once trapped in frozen clathrates beneath the sea, may have been released into the atmosphere and oxidised, stripping away another greenhouse buffer. None of these alone explains a freeze so intense that glaciers spread to the equator. Together, they suggest a cascade of triggers, each feeding into the next, until Earth became a world sealed in ice. A planet that is usually restless—exchanging heat, water, and air through ceaseless movement—stilled. Beneath the frozen crust, magma continued to surge, volcanoes erupted, and gases seeped upward. While the surface world held nothing, the planet's deep machinery continued to provide environments essential for life to cling to the margins.

Descent into Global Ice

The surface of the Snowball Earth was a desert of ice, but thanks to water's molecular properties, beneath the crust the oceans remained liquid. Pressure kept the depths from freezing, and from the seabed rose heat. Hydrothermal vents—gushing with minerals and boiling water—built oases warmer and dark. Around them, microbial mats clung to stone. They consumed sulphides, iron, and methane, weaving chemical gradients into energy.

Closer to the ice, survival took other forms. Salt trapped within the crust created brine channels, thin veins of liquid where freezing points fell. In these corridors, light filtered dimly from above. Photosynthetic algae clung to ice crystals, scattering flecks of green against the white. Bacteria floated in translucent films. Modern Antarctic sea ice holds the same structures, networks of brine that glisten with living cells. They collapse easily, but their presence today offers us a crucial understanding of how ancient life survived here for so long.

Other microbes may have taken refuge in rock pores, where groundwater lingered unfrozen, warmed by geothermal heat. Some slowed their metabolisms almost to stasis, surviving centuries in suspended animation. Others rewired their biochemistry: reinforcing their membranes, altering enzymes to function at lower temperatures, and repairing DNA more often than they replicated it. Life leaned on old tools in new ways, sharpened by necessity.

Multicellularity, only just beginning to emerge, would have faced severe constraint under such conditions. Large, surface-dependent bodies were poorly suited to a planet sealed in ice, and the Cryogenian likely narrowed life back toward microbial-scale refuges. Survival meant reduction and simplification. Life contracted into threads stretched across extreme holdouts. These were corridors narrow enough to have permitted life to fail.

Imagine an ocean roofed by ice, lit only by the magma escaping fissures, sustained by black chimneys rising from the crust. Imagine a planet where silence ruled the surface while hidden mats of bacteria clung to existence. The future of forests and animals, of eyes and skeletons, was held in suspension within brine veins and vent fields. The continuity of life depended on nothing more than persistence at the margins.

The Breaking of Eternal Winter

For tens of millions of years, the Earth remained locked in ice until the same systems that enabled the freeze reversed. Carbon dioxide seeped into the air faster than it could be absorbed. Lacking rain or rivers, weathering stalled, preventing gas from returning to the rock. Gradually, the atmosphere returned.

When the threshold came, it arrived with ferocity. The greenhouse effect surged, and temperatures climbed. The ice that had locked the continents for ages began to crack and collapse. Floods tore across landscapes, while glacial walls fractured, releasing torrents that carved valleys into bedrock. Meltwater surged to the sea in volumes without a modern parallel.

The geology of this thaw is preserved in cap carbonates—pale limestones that drape across the rubble of glacial deposits on every continent. They formed when the thawing oceans, rich in carbon and calcium, precipitated vast blankets of carbonate in chemical shock. These layers are so distinctive that they can be recognised across the globe, signatures of trauma spread evenly over chaos. They are the scars of a fever breaking.

The return of warmth reanimated the oceans. Microbes, long confined to brine cracks and vent fields, spread into open water. Photosynthetic algae bloomed, flooding the seas with oxygen. Energy surged through food chains that had reduced to almost nothing; evolution seized the open niches. Life had passed through a bottleneck, and on the other side was rewarded with the possibilities of expansion.

This thaw was not guaranteed. Had volcanism faltered, had carbon dioxide failed to accumulate rapidly enough, the planet might have remained locked for millions of years longer. If the cryogenic grip had held, evolution's climb toward complexity could have played out differently—and the chain of events that gave rise to Homo sapiens might never have occurred.

The Persistence of Extremes

The Cryogenian era demonstrates that the climate wasn't like a metronome, which consistently swings back to its original position. It can lurch, lock, and remain in states that feel permanent. Ice once reached the equator and stayed there for tens of millions of years. Then, just as abruptly, carbon dioxide overwhelmed the system and the planet thawed.

We are unlikely to repeat this exact fate. The tectonic alignments, the chemistry, the dimmer Sun of that age—you will be pleased to know that these conditions cannot be reproduced in a lifetime or two. But the principle remains; once feedback is set in motion, they run ahead of us. Ice brightens the planet, warming gases accumulate, and seas rise or fall. Systems that seem vast and immutable bend to small pushes given time.

The lesson for us is not that another Snowball Earth is on the horizon, threatening us. The Earth's climate has tipping points, and once they are crossed, the new conditions can persist much longer than the original disturbance. We live in a narrow, peaceful interval between those extremes, but nothing guarantees it will remain.

After Snowball Earth, the planet shifted into new climatic regimes. Oxygen levels climbed, continents reshaped, and ecosystems tested bodies large enough to leave their outlines in stone. However, if we were to delve deeper into the past, beyond these devastating events, we would discover another rarely told chapter. For nearly a billion years, the planet settled into long biological stasis.

27

A Giga-year Gestation

Mesoproterozoic–early Neoproterozoic (≈1.8–0.8 billion years ago)

Billion-Year Bookmarks

Until now, time has stretched before you for tens of millions of years. Ahead, it opens into billions—a scale that dwarfs what we have traced. Following time through this book has already demanded considerable effort, but here it lengthens further. No familiar creatures will lead us. Here, the story is told through isotopes, mineral shadows, and chemical ratios. The path forward is harder, but worth it. It leads to the foundations of everything.

A billion years feels less like a number than a concept—a thousand million years, or one gigayear in the language of geologists. Over that vast span, the planet underwent its own long gestation. Slow interior chemistry and surface change prepared the conditions for life to bloom.

There will be gaps in the evidence here; you already know that is part of the story. But with enough attention, we can read the hints of life long before it leaves anything we would recognise as a body. A framework emerges from signals in the partial record. The conclusions are grounded in what the Earth has kept. Energy flows, shifting chemistry, and subtle feedbacks shaped its seas and skies. What unfolded in this billion-year stretch was the patient work of a planet setting the stage for life's next surge.

The Billion-Year Lull

From roughly 1.8 to 0.8 billion years ago, Earth entered a vast pause. Geologists call it the Boring Billion. It covers the Stenian and Tonian periods, ages that saw long tectonic calm and slow biological change before the Snowball Earth glaciations. The label sounds dismissive, but it captures a truth about the surface.

For almost a billion years, little seemed to happen. Continents fused into one mass, Rodinia, which sprawled across the equator. With mountains worn down and tectonics softening, erosion slowed, and rivers delivered fewer nutrients to the seas. Phosphorus, fuel for photosynthesis, became scarce. At first glance, this long stillness looks like absence. The lull was deceptive. Inside those microbial seas, eukaryotes—cells organised around a single nucleus—refined their inner machinery. Mitochondria pulsed within. Cells linked in tentative clusters, hinting at what was to come. Evolution's tempo was slow, but its momentum endured.

Complexity arose from this sustained equilibrium. Sex arose, reshuffling genes into endless combinations. Endosymbiosis made predator and prey into collaborators. Multicellularity took root. When the stillness broke, life would have the tools it needed at its disposal.

The Mitochondrial Pact

In the seas of the Earth during the Boring Billion, nothing visible to the human eye would have seemed to change. Beneath the surface, evolution was drafting one of its most radical innovations.

Picture a microscopic encounter, magnified until it fills your view. One cell drifts toward another. The larger, an archaeon—a microbe with odd, flexible membranes, a survivor of environments too harsh for most. The smaller, a bacterium is efficient, fast, and skilled at harnessing oxygen. Under normal circumstances, one would consume the other. The archaeon would engulf its neighbour, strip it for parts, and continue on unchanged. But on one forgotten day, that did not happen.

The bacterium was taken inside, but it was not digested. Instead, it remained intact, alive with energy, its chemistry undisturbed. The archaeon divided, and the passenger went with it. What should have been a meal became an inheritance, and hunger turned into symbiosis.

This was endosymbiosis: a symbiotic relationship in which one cell dwelt within another, each changed and moulded by their proximity. Over time, the two struck a bargain. The archaeon provided shelter and access to resources. The bacterium excelled at burning oxygen, its enzymes turning molecules into surges of usable energy. Gradually, the bacterium lost its independence, surrendering genes to its host, paring itself down to a single function. It became a mitochondrion, the energy organelle that still sits inside every eukaryotic cell today.

The transformation unfolded over millions of years, with many false starts and dead ends. Once established, it rewrote the rules of biology. Mitochondria generate energy with unprecedented efficiency. They freed their hosts from the metabolic constraints of prokaryotes. A lone bacterium could only sustain a limited genome, dividing fast and surviving on speed. A eukaryote, powered by mitochondria, could grow larger, stockpile genes, and explore cellular complexity.

It is easy to underestimate the radical nature of this. For over two billion years, life had been simple: single cells without compartments, reproducing by splitting in two. With mitochondria, cells built inner systems. They shaped cytoskeletons to hold form, endoplasmic reticulum to fold proteins, and nuclei to guard DNA. Each eukaryote was a city compared to the huts of prokaryotes.

What factors contributed to the timing of this merger? The particulars are still shrouded in mystery. Perhaps the archaeon failed to digest its prey. Maybe the oxygen-rich world created stress for the host. The bacterium thrived in it. Whatever the spark, the outcome was lasting. Two lineages that had once been rivals now moved forward as one.

This moment is sometimes called the most important event in life's history since life's origin. Without it, complex organisms would never have gained the energetic surplus to experiment with form. Muscles, nerves, skeletons, even thought—all depend on mitochondria churning inside our cells. Every heartbeat, memory, and breath is powered by that ancient pact.

The fossil record does not directly capture this transformation. We cannot split a rock and hope to find the first mitochondrion, petrified in time. But the genetic record keeps the imprint. Mitochondria keep fragments of their own DNA, relics of their bacterial ancestry. Their genes are nested within ours, proof of the merger. In every living creature that carries them—plants, fungi, animals, ourselves—the ancient union operates inside each cell.

Seen against the scale of the Boring Billion, this merger is almost invisible. The continents sat still, the atmosphere hovered, the seas stratified. Within that apparent monotony, life performed an act of audacity. It turned predator and prey into partners, rewriting the cat-and-mouse dynamic into collaboration. They unlocked the ability to step beyond the simplicity of prokaryotes into a new domain of possibilities.

Multicellularity

In the Tonian seas, the long rehearsal changed key. Eukaryotic cells, already enlarged by mitochondria and invigorated by sex, explored new ways of living together. Some stayed linked after dividing, forming chains. Others clustered into thin films that congealed into units.

It was the beginning of multicellularity. These were not yet animals or plants, but colonies, filaments, and fragile sheets adrift in shallow water. But their principle was radical: individuality could be sacrificed for the survival of the group. The genome was no longer the possession of a single cell but a community, preserved across many lives at once.

From a planetary distance, nothing appeared to change. Oxygen hovered at low levels, enough to support these modest experiments but too meagre to ignite further. The Boring Billion held its course. Yet in the deep chemistry of the seas, pressure was mounting. Photosynthetic microbes had been spilling oxygen into the water for ages, but trace metal scarcity and feedback loops kept it confined. By the close of this billion-year lull, those balances frayed.

Nutrients trickled back as continents shifted. Oxygen crept upward, lingering longer in the upper ocean. The equilibrium that had restrained life's reach gave way. The world was primed for a break.

Only in retrospect does the end of the Boring Billion carry the weight of a silence before the first note of a biological symphony. Though faint, multicellular life was unmistakably there. Oxygen was stirring, shifting chemistry and pressing on evolution. Soon, the balance would snap. The Earth would lurch into new extremes—seas freezing, continents fracturing, oxygen surging into the atmosphere. The calm would collapse, and with it, the long rehearsal would end.

The Boring Billion gave cells the time to refine themselves, to test cooperation, to discover sex, to absorb mitochondria. When the planet finally shifted, that possibility would surge into view. The silence of these billions of years is deceptive, but essential. Without it, the complexity to come would have had nothing to work upon.

The Invention of Sex

For three billion years, reproduction was mechanical and predictable. A cell grew, copied itself, and split in two; division and duplication. It was a process that required neither courtship nor risk. Bacteria and archaea spread across the planet on the strength of it, multiplying in lava fields and hydrothermal vents, drifting in the upper ocean. Cloning was reliable, relentless, and for ages it carried life, but sameness has its price. Clones share the same strength, and with it the same weaknesses. When a lineage is uniform, a single change in chemistry or a single parasite or virus can threaten to erase it wholesale. Mutation offered variation, but most changes led nowhere, and many damaged what already worked.

In the stillness of the Boring Billion, eukaryotes discovered a different gamble. They began exchanging their genes. Instead of splitting into identical copies, two cells would meet, merge briefly, and reshuffle their DNA. The invention exemplified extravagance. Unlike cloning, sex carried no guarantee. A cell had to find a partner; if it failed, its line ended there. Even when successful, the process was costly, requiring energy and time.

Nevertheless, the advantage was staggering. Variation floods into populations. Weaknesses could be masked, strengths amplified, and novelty tested in every generation. Instead of waiting for random mutations, evolution could now sample endless combinations.

Initially, primitive eukaryotes would have briefly fused, allowing chromosome alignment and gene exchange before separating. Some lineages refined the process, evolving dedicated gametes, cells stripped down to their essence of DNA. Over millions of years, the machinery of sexual reproduction hardened into place. Pairing, recombination, division.

Sexual reproduction also altered the tempo of life: in clonal worlds, change crept in tiny increments. In the sexual world, it came in leaps, each union a roll of the dice. Traits that improved survival or attraction expanded quickly. For the first time, chance and choice shaped existence together.

Humans carry this inheritance deeply. Attraction, shame, lust, power, intimacy—all stem from a biological process older than consciousness itself. Our brains dwell on it. Our gods condemn it. Law and culture seek to control it, even though civilisations depend on the very instincts it stirs. However, at its cellular core, sex is nothing more than the reshuffling of genes in order to create more robust offspring.

Sexual reproduction seemed destined to occur, but it was not. Many lineages never embraced it and still thrive solely through cloning. Sex was a risk, one that could have failed. But once established, it proved too potent to abandon. Variation became the key to survival, and the gamble of reproduction by fusion spread across the tree of life.

The Boredom That Built Us

Rodinia sprawled with its mountains ground down, its rivers lacking nutrients. Oceans are stratified into layers: oxygenated above, sulfidic below. The air hovered at low oxygen levels, enough to sustain microbes but not yet to light the fuse of complexity. Despite the Boring Billion's stability, unseen transformations were already underway in the cells that inhabited its waters.

Mitochondria had embedded themselves, turning meals into power. Sex had multiplied possibilities, weaving novelty into every generation. Now, multicellularity was sketching outlines of future bodies. These were life's successful inventions, hidden in plain sight, rehearsed across a billion years.

Photosynthetic microbes had been exhaling oxygen for ages, yet minerals and sulphides in the oceans absorbed it as quickly as it was produced. As this billion-year period concluded, oxygen lingered longer, creeping upward. Rodinia, long and stable, strained at its seams. Rifts would soon tear it apart, raising mountains, carrying nutrients back to the sea.

This boring billion-year time span was the aftershock of an older upheaval. Over two billion years earlier, photosynthetic microbes had flooded the atmosphere with oxygen, altering ocean chemistry and the very surface of the planet. Our path now leads further back into the epoch when Earth first drew breath.

28

Atlas Oxidised

Paleoproterozoic—Siderian–Rhyacian (≈2.5–2.05 billion years ago)

Take a moment and inhale, then exhale it back out again. With every breath you take, oxygen rushes into your bloodstream and binds to the iron in your blood. The act of breathing underpins every thought, every twitch of a muscle, every heartbeat you will ever have. The average human may take six hundred million breaths across a lifetime, and many never once stop to consider what an anomaly it is to live on a planet where the atmosphere has enabled this process.

For most of the universe, oxygen is either locked into rocks or too reactive to linger. Mars offers only carbon dioxide in its dim sky. Venus is insulated by a thick layer of carbon dioxide and sulphuric acid. The giants farther out hold onto hydrogen, helium, methane, and ammonia. Though oxygen atoms abound in ice and minerals, free, breathable gas does not exist there. To our knowledge, Earth is the only planet that has this attribute. Three billion years ago, however, the atmosphere you breathe would not have existed; it would have been impossible to breathe on our world. In the age when oxygen transformed the sky, it was as if Atlas—the Titan of Greek myth, doomed to bear the heavens—had been given a new and heavier burden, conjured by the work of the microbes below.

The younger Sun, only three-quarters of its present strength, still shone down to the surface. But the warmth it cast struggled to cut through an atmosphere smothered by volcanic gases. Methane, carbon dioxide, sulphur dioxide, and hydrogen poured from fractures in the crust and filled the air. Without oxygen to give the sky its scattering blue, the horizon glowed in iron tints; shades of swamp green and molten orange filled it.

The seas below were no safer. They were heavy with dissolved iron, a metallic broth that swallowed light and gave the water a deep green-brown hue. Each wave that broke on barren shores left behind traces of reddish sediment, blood-coloured streaks from minerals drawn up from the depths. The air would have stunk of sulphur. Ozone had not yet formed to absorb ultraviolet radiation, so each shaft of sunlight carried lethal force. Daylight was a hazard; the surface was a killing ground.

Despite the hostility, life continued its progression. Microbial colonies clung to the scalding margins of hydrothermal vents, where chemical gradients offered a narrow foothold. These were organisms that asked nothing of oxygen. Instead, they drew their energy from hydrogen, methane, iron, or sulphur. Anaerobic microbes—life that flourishes in the absence of oxygen—dominated every corner of the biosphere. Some of these forms would not only find oxygen useless but would be fatally poisoned by it. In their world, oxygen was a threat best avoided, more than a gift waiting to be unlocked.

This was the baseline Earth for more than a billion years; seas that gleamed with metals, skies dense with methane haze, and continents bare and eroding. Life remained simple, invisible, and microbial. It was a self-contained ecosystem of bacteria and archaea, trading chemical currencies we struggle to frame as sufficient for survival. But they did survive; if anything, they thrived.

Oxygen was so scarce that it left only a minimal chemical trace. Rocks formed in these early ages carry evidence of minerals that cannot coexist with free oxygen. Ancient soils preserve iron in its reduced state, unbound and unoxidised. The chemical signatures from these palaeosols, ancient soils locked into rock, show that the atmosphere was suffocated by a lack of oxygen. The world was stable in its strangeness, and nothing suggested that this balance would break. Some microbes had already been harvesting light without oxygen, using sulphur or iron to drive their reactions, but the chemistry of water lay waiting.

The precise tempo and timing of the first oxygenation remain active areas of research. Evidence suggests there may have been multiple pulses or fluctuations in oxygen levels, rather than a single, smooth transition. Some of these shifts coincide with global glaciations, including the Snowball Earth episodes described earlier.

Oxygen's First Ascent

The first innovators were small single cells, no larger than dust motes in the water. The energy they had access to was limited, as iron was only abundant in certain basins, sulphur in vents and seeps. Then, between 2.7 and 2.5 billion years ago, cyanobacteria evolved the biochemical machinery that would enable them to utilise another energy source. They turned their attention to water, the most stable, stubborn molecule in reach.

Splitting water is a complex process, especially against the backdrop of organisms that were little more than drifting cells. Its bonds are tight, and most chemistry on Earth cannot prise them apart. Nevertheless, within the folds of a cyanobacterial membrane, a cluster of manganese and calcium atoms arranged themselves into a tiny catalytic machine. This oxygen-evolving complex, working in tandem with a relay of pigments and proteins known today as Photosystems II and I, could strip electrons from water molecules one by one. Sunlight drove the reaction, water surrendered its charge, and oxygen drifted free as a waste product.

These cells now had an inexhaustible source of electrons. Unlike sulphur or iron, which appeared in scattered areas, water covered the globe. Cyanobacteria bloomed across the shallows. Their colonies layered into stromatolites, sedimentary structures that crept upward from the sea floor. Some of their fossil traces can still be found embedded in rocks in Australia and South Africa.

The oxygen they released was volatile, reactive, and hostile to the world that had existed before. At first, the oceans absorbed it. Iron ions dissolved in seawater bound to it, precipitating as iron oxides that sank to the seafloor. Over millions of years, this reaction formed immense banded iron formations: red and black rock layers that are still visible in regions like the Hamersley Range in Western Australia. These ancient seams, mined today for steel, are evidence of some of the first signs of oxygen.

For a time, the seas kept swallowing the gas, locking it away in stone. But the production of oxygen was relentless once it had been configured. As the oceans' iron stores became saturated with oxygen, whatever could not bind to iron drifted into the sky. There it collided with methane, Earth's previous atmospheric blanket. Methane is a potent greenhouse gas, and in the absence of oxygen, it kept the faint young Sun's warmth close. Once oxygen stripped it away, the planet lost that insulation. The same oxygen that rusted seas was now scouring the sky, reshaping weather, and plunging Earth into cold. The gas wove itself into rock, air, and climate concurrently.

For the anaerobic majority, oxygen was a toxin they had little to no defence against. Cell walls ruptured, enzymes corroded, and DNA frayed under its assault. Where once they had spread across ocean basins, they now retreated into marginal refuges—sulphur vents, hyper-saline pools, deep sediments where oxygen could not penetrate. Some clung on, and descendants of these archaic lineages survive today in the dark recesses of Earth's biosphere. But it was here that they lost their dominion of the planet.

Oxygen gave these early photosynthetic cells energy, while driving countless other lineages to collapse. Microbial diversity contracted; the very air was becoming corrosive. Ancient sediments lose the peculiar fingerprints of sulphur isotopes that had marked a sky absent of oxygen. Rocks shift from containing reduced minerals—those that only survive without oxygen—to oxidised forms that require its presence. Red beds tinged with ferric iron appear on land. These are the signatures of a planet tipping. By 2.32 billion years ago, Earth's atmosphere held a permanent trace of oxygen; the Great Oxygenation Event had begun.

It destroyed most of what lived, but in the wreckage lay the chance for life to grow beyond what was ever possible before.

The Fuel for Complexity

Anaerobic life could ferment sugars or strip electrons from sulphur or iron, but the yields were meagre. A single molecule of glucose broken down in this way returned only a faint ratio of ATP, the energy currency of the cell. With oxygen, the same molecule could yield more than fifteen times that amount. The difference was the gap between subsistence and abundance. Once the process had started, cells could afford to grow larger, maintain more elaborate machinery, and replicate genetic material on a new scale. Oxygen changed the ceiling of what life could become.

This energetic revolution set the groundwork for complexity. Larger genomes demanded a reliable energy supply. More intricate membranes, internal compartments, and eventually organelles like mitochondria depended on it. The potential for multicellularity and for bodies composed of cooperating cells came in the aftermath of oxygen's arrival.

But oxygen was—and remains—ferociously reactive. In its presence, one had to evolve layers to navigate the duality. Antioxidant enzymes, repair pathways, and new metabolic circuits had to develop to contain the damage wrought by reactive oxygen species. Even within us, each breath feeds the brain, but it also breeds free radicals—molecules that chip at DNA, deform proteins, and accumulate as the years advance, contributing to the slow decay we call ageing.

Above, in the skies, with enough oxygen present, some molecules absorbed ultraviolet light and split, only to recombine as ozone. This thin veil of O_3 spread across the upper atmosphere, intercepting radiation that once scythed down unopposed. Life was shielded from the sun for billions of years, living in deep water or under rock. Ozone's protective layer offered a haven on the surface. Shallow waters and coastlines could harbour life without instant genetic ruin. Oxygen had made the world dangerous, but in combination with sunlight, it was now also making it more hospitable.

It's easy to oversimplify this process; oxygen levels would have continuously surged, fallen back, and surged again. For hundreds of millions of years after the initial oxygenation, concentrations saw-sawed, sometimes near modern levels, sometimes collapsing to a fraction of a per cent. The history of oxygenation is an ongoing subject of scientific debate—geological and chemical signals suggest complex fluctuations, and many details about the duration and extent of these changes are still being resolved. But given the time frame, the general direction still held upward. Oxygen became part of the fabric of Earth's metabolism.

Today, the balance rests at just under twenty-one per cent. It is high enough to fuel complex life but not so high as to let flames consume the biosphere unchecked. This ratio is precarious, held by the churn of ecosystems—plants releasing oxygen, microbes consuming it, fires and decay pulling it back down. Earth acts as one vast system, and oxygen is its pulse. If oxygen fell too low today, breath would falter, muscles would seize, and consciousness would fade into the dark. A fraction higher, and lightning would do more than just strike single trees—it would ignite entire continents.

This planet burns just slowly enough not to catch fire.

That balance isn't fixed, and there's no certainty it will stay the same. Oxygen is fuel and toxin, both shield and axe. It lifted the ceiling of energy and made room for complexity, but it demands constant negotiation. It allows the surface to host forests, seas filled with animals, and skies filled with birds.

Breathing Hazards

Individual O_2 molecules turn over in the atmosphere within thousands of years, destroyed and remade in cycles of photosynthesis, respiration, and fire. The oxygen atoms feeding your brain now may once have rusted into banded iron on the Archaean seafloor, been freed again by weathering, carried through the leaves of Carboniferous forests, or bound to the blood flowing through the vessels of a Jurassic sauropod.

In the oceans, cyanobacteria and plankton pump oxygen back into the sky each day. On land, forests do the same with their leaves. Animals draw it down and give it back as carbon dioxide, which plants and microbes reconfigure again into sugars and more oxygen. Fires consume it in violent bursts; decay uses it more slowly. Soil microbes, fungi, roots, and rivers all contribute. The balance of twenty-one per cent is held by this churn.

For much of Earth's history, oxygen has not been abundant enough to permit animals of our size. It is only in the past six hundred million years that concentrations stabilised high enough to allow bodies with nerves, muscles, and skeletons. As we saw in the Carboniferous, oxygen spiked to extraordinary levels, fuelling dragonflies with wingspans the length of hawks and wet forests that burned as if damp with petroleum. Later, it dropped, threatening global suffocation. Oxygen has remained stable throughout human evolution within the narrow band where fire can shape forests without consuming them, and lungs can draw breath without collapsing. That this balance holds at all is among the rarest contingencies in Earth's story. We are accustomed to thinking of oxygen as abundant, given, and inexhaustible. However, it results from numerous factors coming together. Every lungful depends on the resilience of forests, the bloom of plankton, and the integrity of soils.

There is something humbling in this recognition. The Great Oxygenation Event was a byproduct of microbes searching for energy. They released a gas that corroded their neighbours and nearly froze the world. Yet, from that unintended shift came the possibility of complex life. The most dangerous molecule in the early biosphere became the foundation of consciousness.

So take another breath, feel your lungs expand. Oxygen binds to haemoglobin, and the heart pumps it forward. You are enacting a ritual that links you to stromatolites older than mountains, to rusting seas; the same element that erodes your cells also ignites your thoughts.

Oxygen began as a poison. It remains dangerous still. But it is our poison, and it has made us everything we are.

29

Anaerobic Supremacy

Archaean Eon (≈4.0–2.5 billion years ago)

By the time the Archaean Eon began, over four billion years ago, life may have already been established. The planet had not long emerged from the Late Heavy Bombardment, a storm of asteroids and comets that scarred its surface and boiled its oceans. The crust was incipient, unsettled, and volcanic. There were no continents yet, only scattered volcanic islands engulfed by acidic seas. Days spun faster—perhaps only fifteen or sixteen hours long—and the Moon loomed far larger, dragging tides that surged hundreds of metres across barren shores. The Sun, as before, was at only three-quarters of its current strength. Without a dense greenhouse blanket, the oceans would have frozen.

The oldest convincing traces of biology come from rocks over 3.7 billion years old in Greenland's Isua supracrustal belt. Within them are carbon isotope ratios skewed in ways consistent with biological processing, and textures that resemble stromatolites. The evidence is debated, but if true, it implies that life arose astonishingly soon after Earth became even marginally habitable. In planetary terms, biology was strangely almost immediate, woven into the crust before the surface had fully stabilised.

The first organisms were prokaryotes—cells without nuclei. They drew power from the planet's raw materials. Some consumed hydrogen and carbon dioxide, exhaling methane into the air. Others fed on sulphur compounds or iron ions, driving their metabolisms the way we drive ours with oxygen. Each lineage was tuned to the gradients of heat, minerals, and gases in its surroundings.

Anaerobes spread across every environment the young planet offered. They lined the seabed in dense films, colonised volcanic margins, and drifted in shallows where sunlight met mineral-rich waters. They grew into mats that trapped sediment and bubbles, layering themselves into structures that puckered the shoreline. Some seeped into the crust itself, kilometres down, where fractures carried hot fluids through basalt. Their metabolisms cycled waste into food, closing loops so that little was lost. For more than two billion years, this was the biosphere: microbial and self-contained.

It was Earth's first empire.

Anaerobes endured glaciations, comet strikes, volcanic winters, and climate shifts that would later erase entire animal lineages. They survived everything the planet hurled at them. For sheer longevity and planetary influence, the reign of anaerobic microbes dwarfs anything life has achieved since.

Stromatolites, the most visible monuments of this reign, still rise in the rock record. These domes and ridges trapped sediments that are preserved in ancient strata from Australia, South Africa, and Canada. Some still grow today in a few protected lagoons, layered by cyanobacteria and their microbial companions.

The reign of the anaerobes did not end with the rise of oxygen. Even today, they still account for a vast portion of Earth's biomass. The deep biosphere beneath our feet—microbes living in aquifers, sediments, and fractured rock kilometres down—may contain hundreds of billions of tonnes of carbon, possibly rivalling or even exceeding all the forests and animals at the surface, though estimates remain uncertain. These cells live in darkness, often at high pressure and temperature, and some divide once in millennia. They are invisible to us, but they outnumber us, outweigh us, and will almost certainly outlast us.

Their persistence unsettles our sense of destiny. We imagine intelligence and complexity as the natural outcomes of evolution, as if the universe had been waiting for minds. But the weight of history lies elsewhere. For half the planet's span, life flourished without eyes, nerves, or thought. By any measure of continuity, this empire is the most successful in Earth's history.

Organic Clues from the Void

From orbit, Earth would have looked dead: no bio-signatures to read. The seas were alive with microbes, but the sky gave away nothing. For two billion years, the planet supported a biosphere that left no signal detectable from afar. Anaerobic metabolism is stripped-down and adaptable. They do not need oxygen or even sunlight. Some utilise hydrogen bubbling from fractured rocks. Others reduce sulphur compounds seeping from vents. Still others use iron ions dissolved in seawater as electron acceptors. They thrive in darkness, in sediments, and in aquifers kilometres down. Their cycles are self-contained: waste from one cell becomes food for another.

This invisibility matters because most of our search for life elsewhere is built on the opposite assumption. Telescopes are tuned to detect atmospheric biosignatures: oxygen, carbon dioxide, and methane in unstable combinations. The James Webb Space Telescope and the missions that will follow it scan distant exoplanets for chemical imbalances that suggest the presence of life. On Earth today, oxygen's presence in the atmosphere is a glaring signal of life. But for most of Earth's history, that signal was absent. If another civilisation had been watching, they might have dismissed our world as a sterile rock, as we have done with countless others.

That warning sharpens as we look at the planets closest to us. Europa and Enceladus hide salty oceans beneath ice crusts tens of kilometres thick. Plumes jet from their surfaces, rich in organic molecules. Inside, water meets stone at hydrothermal vents, a setting remarkably like the one that hosted Earth's first microbes. If life thrives there, it is almost certainly anaerobic. We could orbit those moons for centuries, scanning skies that tell us nothing, while whole biospheres work unseen in the dark below.

Mars may still hold a shadow of such life. Its surface is harsh—frozen, irradiated, stripped of air—but deep underground, water lingers. Where water seeps into basalt, hydrogen is released. That chemistry is as steady as the tide, and enough to sustain microbes if they exist.

Meteorites further reinforce this possibility. In 1969, a fireball exploded over Murchison, Australia, scattering fragments of carbon-rich rock. Inside were over seventy kinds of amino acids, along with nucleobases and simple sugars. None were contaminants; they had formed in interstellar space, frozen into a parent body and delivered intact to Earth. Comets show similar chemistry. The Rosetta mission to comet 67P detected glycine, a building block of proteins. Similar asteroids rain down on Earth every day in the form of dust and rock, as they have for billions of years. The cosmos may be more generous with life's ingredients than we have ever been led to believe.

If the building blocks are common, then the first metabolisms may also be common. Many anaerobic pathways are chemically straightforward compared to oxygenic photosynthesis, though some—such as sulphur and iron metabolisms—involve specialized enzymes and complex biochemistry. Hydrogen metabolism, sulphur reduction, and iron respiration—these reactions often require little more than a cell membrane and a few catalytic proteins. They do not alter a planet's atmosphere.

This raises a sobering possibility. The galaxy may be filled with living worlds that appear sterile. Planets could host oceans thick with microbes, crusts riddled with cells, and atmospheres tuned by chemistry but never tipped by oxygen. From light-years away, they would look like bare stone. We would miss them, just as another civilisation might once have missed ours.

For most of Earth's history, life moved in silence, a microbial world without breath or sound. Oxygen broke the quiet only once, as far as we know, in the Great Oxidation Event. It was no foregone destiny but a rare conjunction of biology and geology—a fluke whose rarity may mean the common face of life in the cosmos is the one that leaves little trace.

It is humbling to imagine the galaxy filled with such worlds. Oceans sealed under ice, crusts lined with microbes, rogue planets humming with chemistry in the dark. They would never draw attention to themselves, never flare into visibility.

Once cyanobacteria learned to prise electrons from water, the atmosphere changed, rocks rusted, and the sky turned blue. From that point on, Earth would look alive even from the stars. But for two billion years, it did not.

The Microbial Circuitry of a Young World

The oceans of the Archaean were propellants, wired directly into the restless geology of the planet. Their waters gleamed with dissolved iron, their depths burped with volcanic gases, and plumes of scalding water jetted from the seafloor. Into this turbulence, microbes wove themselves into the circuitry. They turned chemical gradients into metabolism, and together they kept the planet alive in silence.

Along the shallows, where light could filter through the haze, some cells bent sunlight to their advantage, though not in the way familiar to us. They did not prise oxygen from water. Instead, they stripped electrons from ferrous iron dissolved in seawater. The waste dropped as ferric oxides, staining the seabed red. Over immense stretches of time, these precipitates stacked into layered cliffs of black and crimson: banded iron formations. They still stand today in the Hamersley and Pilbara ranges, stone diaries of oceans that rusted without air. When we trace their ridges, we're essentially feeling the pages of ancient microbial chemistry, dating back three billion years; every stripe represents cells that captured light in a time when there was no oxygen.

Deeper down, beyond the reach of day, other organisms survived on stipends of hydrogen. Water locked in minerals was split by radiation or by the reaction of hot fluids with iron-rich rock, a process known as serpentinisation. Gas leaked into aquifers and fractures. For microbes sealed in the crust, this was enough.

At the blurred frontier between sediment and sea, methanogens stitched hydrogen and carbon dioxide into methane. They filled the atmosphere with a greenhouse gas that retained heat from the younger Sun. Without them, the oceans might have frozen, smothering the biosphere beneath ice. For millions of years, they maintained the planet's rhythm, their power tied to metals flowing from the mantle. Their enzymes depended on nickel, once abundant but gradually dwindling as Earth cooled.

Even methane was not free to accumulate without limit. In anoxic muds, microbial partnerships evolved to consume it without oxygen. One partner split the molecule; another bound the released electrons into iron or sulphur compounds. These collaborations, invisible on the surface, acted as brakes, reining in the greenhouse blanket before it ran out of control. Together, these pathways formed a hidden grid. Iron oxidisers rusted seas with light. Hydrogen feeders survived on stipends from the crust. Methanogens exhaled a blanket of warmth, while consortia consumed the excess in the dark. Each metabolism fed into the next. Waste became food, loops closed, and almost nothing was lost.

Banded iron cliffs testify to the seas where light and iron met. Sulphur isotopes in ancient rocks carry the chemical fingerprints of skies without oxygen. The eventual decline of these iron formations around 1.8 billion years ago marks the intrusion of oxygen, when the old currency of the oceans lost its value.

This hidden grid was Earth's first organic stability. Compared to what followed—the oxygen revolution, the Cambrian burst, the rise of forests—it was a world without spectacle, so muted it leaves us relying on abstraction to imagine it.

The Archaean oceans show that a biosphere can endure without reshaping its planet. In its simplest forms, it rode the planet's shifting climates with unshaken stability. Before those pathways began, Earth held only brief chemical flares—reactions that sparked and guttered without memory. Then, somewhere in that turbulence, a loop closed: molecules learned to keep their own pattern, to feed, to divide—matter crossing the threshold from the inorganic to the living.

30

Ex Nihilo

Hadean–early Archaean (≈4.6–3.5 billion years ago)

Lightning in a Bottle

In 1952, in the basement of the University of Chicago's chemistry department, a young graduate student named Stanley Miller assembled what looked like a storm trapped in glass. Working under Nobel laureate Harold Urey, he connected a chain of flasks and tubes, filling one with water and heating it to mimic a primitive ocean. The vapour was drawn upward into a second chamber that contained a mixture of methane, ammonia, and hydrogen gases thought to resemble Earth's early atmosphere.

At the top of this chamber, Miller installed two electrodes. They spat continuous sparks to represent lightning, flashing across the gases day and night. Steam rose, mixed with the atmosphere, was struck by an electric charge, condensed, and fell again into the flask below. This was a closed cycle that replayed itself. At the time, it was widely believed that Earth's first atmosphere was rich in methane and ammonia, a chemical playground for the formation of new molecules.

Since then, the picture has grown more complicated. New evidence suggests the air might have been less friendly to such easy chemistry—more carbon dioxide and nitrogen, less of the ingredients Miller used. But the point remained: even in a tougher environment, sparks and water could drive matter toward complexity. The details have evolved, but the experiment's lesson holds true—a young world can coax order out of storm and stone.

After only a few days, the clear liquid inside the flasks went from translucent to a cloudy opaque mixture, then turned the colour of weak blood. When Miller analysed the darkened broth, he discovered molecules that had not been present at the start: amino acids such as glycine, alanine, and aspartic acid, the same monomers that living cells string together into proteins.

Life was not created here, but it was an incredible breakthrough nonetheless. From nothing more than gas, water and lightning, matter had been driven into patterns more complex than chemistry alone was expected to yield. It turned out that the boundary between the inert and the biological was thinner than many had previously believed.

The demonstration was not infallible proof of life's origin, but it revealed a principle that altered the conversation. The essential components of biology could have emerged naturally under the conditions available on early Earth. What Miller recreated in a basement might once have happened in tidal shallows, volcanic skies, or storm-lashed pools. Each strike of lightning could have left residues—most of them ephemeral, some enduring long enough to meet and combine—until feedback loops formed and persistence began.

From the reddened broth of that flask, we turn to a world still molten and unstable. Lightning lashes oceans that have only just gathered from steam; skies writhe with methane haze; even continents don't exist here. What Miller glimpsed in days, the planet rehearsed for aeons. Storms write alphabets in water, chemistry leaning toward continuity. Out of that turbulence, patterns clung, and from their tenacity, the history of all living beings would begin.

Theia

The Hadean is the earliest eon of Earth's history, lasting from 4.54 to 4.0 billion years ago. The Solar System was still young. Planets were coalescing from collisions measured in millions of years. Earth took shape, a growing sphere still hot from its own assembly. The evidence suggests that not long after, a planet the size of Mars struck it at an oblique angle. The collision tore away part of Earth's mantle and threw debris into orbit, where it gathered into the Moon. Without the Moon, Earth's tilt would have wandered chaotically, its climate would have remained unstable, and its seas would have been unpredictable. Already, chance collisions were setting the conditions that would favour the genesis of life. Ex nihilo—'out of nothing'—is the old language of creation, a human phrase laid over a world assembling matter, heat, water, and time.

Molten rock surged like glowing tidal surges across the crust. Planetary debris rained down, sometimes smaller than mountains, sometimes large enough to boil seas and reset the atmosphere. For a long time, scientists believed in an episode called the Late Heavy Bombardment—a spike in cosmic violence that scarred the young Earth and erased most of what it briefly built. Now, some wonder if the barrage was less a sudden strike and more a drawn-out hail, falling steadily over hundreds of millions of years. Either way, it was a world under siege, its surface built and unbuilt again and again. In the intervals between impacts, lava cooled into rafts of crust, only to be shattered again. Volcanoes poured silicate vapour into the air. The sky filled with ash, opaque to starlight. Days were shorter, the planet spun faster, and the Moon's pull dragged tides across basins with colossal heave.

What could endure in such a tumultuous world? We have no fossils or sedimentary rocks to point to. In the Jack Hills of Western Australia, tiny zircon crystals over four billion years old contain isotopic ratios that suggest the presence of liquid water on the surface. They are minuscule, grains no larger than a pinhead, but they tell us that even amidst fire and bombardment, rain may have fallen, and seas may have gathered in the hollows of a cooling crust. These zircons are among the oldest objects you can hold in your hand, fragments of an Earth still half-formed.

The name Hadean is well chosen. In Greek thought, Hades ruled the unseen, not death, the shadowed domain beneath all things. The Earth of this eon was exactly that: a foundation hidden beneath everything, mostly unknowable, yet ultimately essential to the story. From its fires came the crust we walk upon, the field that shields us from the solar wind, the Moon that steadies our tilt, and perhaps even the first drops of water that would one day enable life.

The surface of the Earth was temporary, because the forces at work erased as much as they created. For hundreds of millions of years, the planet held no permanence, only the flux of tectonics and impacts from above. A magnetic field formed as the core churned. Oceans condensed, however briefly, on stone hot enough to steam. An environment completely hostile to us, but one that contained the seeds of everything we depend on. The first ingredients for life—water, energy, chemistry—were assembling in a world that seemed incapable of holding them.

The Making of the Hydrosphere

The rain would have hammered down in tempests, torrents struck bare rock, breaking into a sibilance of steam and spray on contact with molten plains. Lightning flashed in the clouds, thunder echoed across oceans, and vapour collapsed into floods that collected in hollows and basins. Oceans accumulated until a world of fire had to share its home with water.

The Zircon crystals carry oxygen-isotope ratios that indicate liquid water existed by 4.3 billion years ago. That is astonishingly soon, only a geological heartbeat after the Late Heavy Bombardment afflicted the surface. Where it all came from is still a matter of discovery and debate. Some of it was intrinsic, bound into silicate minerals at Earth's birth, driven outward as steam through fractures and volcanoes. Some was delivered from beyond, by asteroids loaded with hydrated minerals. Hydrogen isotope ratios in modern seawater point to this dual heritage—water from within, water from without, mingled into one ocean.

More improbable than its arrival was its survival. Earth stayed wet, but the balance keeping it wet was narrow. Strong enough gravity to hold vapour, a magnetic field to guard the atmosphere, a Moon to stabilise axial tilt, and just enough greenhouse gases to keep the surface from freezing. Even then, the ocean could easily have been lost. Recycling between the crust and the mantle kept it alive. Slabs subducting into the deep carried water down; the planet became a reservoir for it.

Today, the hydrosphere holds 1.4 billion cubic kilometres of water, enough to cover the planet to a depth of three kilometres if evenly distributed. Still more lies hidden in mantle minerals, locked under crushing pressures—maybe even as much as in all the surface oceans, or perhaps much less; the true scale is still being uncovered. The Earth is threaded with water, in rivers and glaciers, in crustal pores, in clouds and storms, in the slow churn of subduction and volcanism. Once liquid, water became an agent of change. It weathered rock, carried minerals, dissolved and recombined elements. Its polarity—the slight imbalance of charge across the molecule—made it a universal solvent, able to coax atoms into interaction that air and stone alone could never achieve.

This is why every search for life begins with the same question: Is there water? Without it, molecules remain locked in place, chemistry inert. With it, reactions cascade, and loops close. Had the mantle cooled differently, or the orbit tilted closer to the Sun, or the collisions delivered less, the oceans would not be here. Their continued existence is among the more unusual circumstances in this story.

In those waters, atoms drifted and met, held for a moment, then slipped apart. Chains formed and broke, while water made them linger. It folded chance into repetition, driving the same collisions through pools that dried and refilled, through vent walls plated with minerals, through storms that raged and dispersed. What was scattered in the air and stone was gathered in liquid, pressed together until the improbable looked inevitable.

The question still burns: in that chaos, by what hand of chemistry did life begin?

Origins of Life

The mystery of life's beginnings has never loosened its grip on our collective consciousness. It is among the most studied questions in science and philosophy, and still one of the most incomplete. We do not know whether the first spark was caught in the glare of the Sun or in the darkness of the abyss. We do not know whether it unfolded in the rhythm of tides or in the fissure of stone. Perhaps it was sudden, a cascade of reactions that locked into self-sustaining loops in a geological instant. More likely, it was drawn out, a slow accretion of chemical habits refined across millions of years. We know only this: it happened once, and everything else is a consequence of that event. For all our experiments and equations, no single vision yet holds the answers. Instead, we stand before a puzzle, each piece offering clues, but none complete the story on its own.

One trail begins at the bottom of the ocean, where the planet's restlessness splits the crust and pours heat into water. Hydrothermal vents—black smokers and alkaline chimneys—still climb from the seabed today, their spires built from minerals precipitated as hot fluids meet cold seawater. Within their porous walls, gradients of temperature and charge separate the two sides. Here, carbon dioxide dissolves, iron and nickel catalyse reactions, and simple molecules are coaxed into chains. These are natural reactors that have operated continuously for millions of years. Nick Lane and others argue that it is here, in such places, that metabolism first arose. Networks of reactions driven by energy differentials, capable of feeding themselves. The power of this idea lies in its continuity. Every cell alive today runs on proton gradients across membranes.

But there are doubts about this theory. Can vent chemistry move beyond scaffolding reactions to the intricacy of replication? Can combinations of carbon and hydrogen, no matter how elaborate, ever bridge the gulf to heredity? Critics point to the need for a molecule that can record, to carry information forward; that brings us to another trail.

The primordial soup, as it came to be known, imagines a different stage. Shallow pools under a stormy sky, warmed by volcanic heat and stirred by lightning. The soup is romantic in its simplicity, lightning striking puddles to create the alphabet of life. Miller and Urey built their glass flasks to test this vision. However, Early Earth's atmosphere may not have matched Miller's recipe. Amino acids alone do not organise into cells. Sparks make organic molecules, but an enormous piece is missing.

Others turn to RNA. Unlike DNA, which simply stores, or proteins, which simply act, RNA can do both. It can hold sequences of information and fold into catalytic shapes. Laboratory experiments have produced ribozymes that snip, join, and even copy other RNA fragments. The RNA world hypothesis imagines a time when such molecules ruled alone, copying themselves with errors, mutating into diversity. It is compelling because it unites heredity and function, but it too is precarious. RNA is chemically unstable; its nucleotides are hard to assemble under natural conditions. And despite decades of effort, no one has built a fully self-sustaining RNA system without intervention. It's plausible enough to have existed, but perhaps not as the first step.

And then there is the most vertiginous possibility, that life did not begin here at all. Panspermia—the cosmic seeding of biology—suggests that microbes or their precursors arrived aboard meteorites or comets. We know amino acids and nucleobases pepper some meteorites. We know hardy spores can survive the vacuum of space. Mars, with its once-warm surface and rivers, is often named as a possible cradle, with Earth receiving its gift after an impact flung debris between worlds. This idea explains little, only moving the question outward, but it lingers because it is not impossible to conceive.

These theories are not mutually exclusive. Each may hold varying fragments of the truth. Life might have required the energy gradients of vents, the sparks of storms, the memory of RNA, the shelter of lipids, and perhaps even the raw material delivered from space. No fossils mark the moment, no sedimentary layers preserve the first loops of carbon or the first accidental replication. All that remains are isotopic threads and the redundancies buried in our biochemistry.

What they tell us is this: life is, at root, a system for remembering. Molecules that once formed and then vanished began, at some threshold, to leave traces of themselves. They coped with variation, endured through cycles of destruction and renewal, and tilted the future toward their own persistence. And it is this memory, encoded and re-encoded across billions of years, that underlies the very thought by which you read this sentence.

But where is the line? At what moment did chemistry lean into biology? Was there a single instant when the inanimate stirred into the animate, or was the shift so drawn out that the boundary dissolved into illusion, a mountain to climb mistaken for a gate to pass through? At a deeper level, it forces us to ask what life truly is: is it a repeating pattern etched in matter, a current of energy held in motion, or something that only appears when a planet's cadences align across time and space?

If life emerges easily—if water, energy, and carbon are enough—then the cosmos may be littered with hidden kin, each running its own experiments. If it does not, if the choreography is narrow, then our oceans may be singular, the only ones ever to have carried the transition through. Either possibility is difficult to grapple with. Either way, we are forced to contemplate why the universe tolerates life at all.

What we can confirm with certainty, though, is that somewhere in that cascade a molecule learned to last, then to copy, then to vary. Other than that, we arrive at the brink of the unknowable. We have not solved the mystery of life's origin, and it's possible we never will. But even the asking is its own revelation: a species consisting of molecules, bound into minds, reflecting on the moment molecules first became themselves.

The Ancestor We All Inherit

Somewhere between magma and water, at the seam of the Hadean and the Archaean, a system of molecules learned to continue forward alongside time. The first pattern in our lineage that refused to vanish. Out of restless chemistry came continuity. If it happened only once, once was clearly enough.

We call that ancestor LUCA—the Last Universal Common Ancestor. Not the first cell, but the last point at which all life came from one. Every root of the tree of life runs back to it. Other experiments may have sparked and failed, but LUCA's memory still runs through everything alive today.

The organism was not recognisable in any sense. It had no eyes, no nerves, no body plan. It bore a genetic code—already recognisable as the same triplet alphabet we use now. LUCA's existence is traced in the echoes inside every living cell. How energy is carried by ATP. The shared machinery of transcription and translation. The proton gradients that hum across membranes, identical in principle to those that flow through your mitochondria this instant.

Think of the improbability. One lineage, among perhaps many, found enough resilience to last. It had no purpose beyond surviving, no plan beyond replication. From that blind continuity came every wingbeat, every thought, every extinction. A lattice of molecules in darkness, already holding the concept of heredity, already capable of error and therefore of change.

We cannot know where LUCA lived, or whether it lived alone. We know only this: had it failed, the chain would have snapped, and the story would have ended before it began. From a handful of molecules in an unforgiving sea, a self-sustaining pattern took hold and continued through every divergence and contraction of the tree of life. Every breath and word, every extinction and renewal, traces back through that descent, across billions of years to the present, when life has become conscious enough to recognise its own improbable existence.

31

Witnesses to the Eleventh Hour

Holocene (c. 11,700 years ago–present)

This is the shape the planet's story takes, as far as we can trace it. Not an exhaustive account by any means, but a scaffold of understanding. Life began in a place the Earth still remembers. Warmth moved through stone, water carried minerals along fractures, and energy spilled over surfaces that coaxed molecules into fragile order. Strands of code copied with mistakes, and those mistakes became novelty. Whether this drama unfolded in submarine vents or tidal flats remains unsettled. The stage is debated; the play is not.

The story of Earth stretches back roughly four billion years, with the Precambrian era accounting for nearly 90 per cent of the planet's existence. That stretch is so vast that it renders other numbers irrelevant. Continents drifted and collided. Seas filled and drained. The climate shifted between furnace and ice. For almost all that time, the world was microbial. Bacteria and archaea reshaped air and stone. Complexity came late, and in stops and starts: a stagger, a collapse, a return. We bear the scars and the successes. Measured against that immensity, our own tenure is thin. Walking apes with tool-ready hands appeared only in the last few million years. The form recognisable as modern *Homo sapiens* has existed for only a few hundred thousand years. Barely three hundred millennia ago, a small hominin lineage appeared.

The phrase 'eleventh hour' originates in the Parable of the Workers in the Vineyard in Matthew 20:9, where workers hired at the eleventh hour receive a denarius, a Roman coin. In it, workers hired at day's end receive the same reward as those who toiled from dawn. The phrase means a late arrival can transform everything. Humanity is that late arrival. Four billion years into life's story, our species entered almost at closing time. We forced the planet onto a new trajectory within a geological instant.

Pandora's Primate

We learned to look both behind and ahead, into pasts we did not experience and futures we will never see. Call it self-awareness, or simply the recognition that a life is never sealed. Each person carried kin, unborn children, absent allies, and the voices of the dead. Being human meant managing time, living now, but planning with memories based on what came before. We organised the next meal, the next season, the next storm, the next birth.

Every lineage faces the same two imperatives: keep living, leave replacements. By 'imperatives,' I mean the basic biological drives to survive and reproduce. Our bodies still answer them. We eat, we avoid harm, we raise children. Yet upon that foundation, we built new demands—the extensions of a creature that learned to negotiate with time. Ritual, art, science, religion, law: different tactics for making meaning endure long enough to matter. A song can gather a people faster than a charter. A charter can hold them together longer than a song. At our best, we braid them and abandon the labels.

Our inventiveness brought both benefits and complications. Growing grain allowed cities to thrive, but also tied them to the success of harvests. Steam engines let people travel far, but they filled the air with pollution. Communication networks extend our reach in ways past rulers could only imagine, but they also distract us to the point that it now feels difficult to focus on a film without checking our phones.

Plastics fixed previous issues, but ended up polluting oceans with waste. Antibiotics beat old diseases but led to stronger, resistant ones. Algorithms now create maps, give medical advice, and make art quickly, but they also shape politics, heighten fear, and divide people into groups based on beliefs rather than family ties. Human attention has become a valuable commodity, often exploited by technology companies. Every invention can help or harm, offering solutions and creating problems at the same time.

There is no return to the firelit plateau. No amount of 'ancestral living' or consumption of raw liver will reverse the journey our species is on. The way forward is not refusal but continuous refinement. Judgement, practised haltingly, has always steered us. Survival at this scale is a matter of choosing what to prize, what to restrain, and what to carry forward despite the cost.

Why This Story Holds

You have followed a thread through stone, isotopes, code, and bones. Here, trust does not mean faith in an author. Mistakes in this account will not last long. They will be found, challenged, and corrected. Claims endure only if they can withstand repetition, boredom, and doubt. New evidence redraws the picture—mostly in detail and rarely in revolution. Plate tectonics was once dismissed as fantasy. Now it is taught to children. Birds were once thought to be separate from their Mesozoic ancestors. Now we know they are living dinosaurs, closer to Velociraptor than to any pterosaur.

Mud hardened into a seabed. It now shapes the calcium in your bones. Iron formed in ancient stars. It flows in your blood with every breath. We can trace how this happened. We can kneel at a cliff and read shells in limestone. Then, we can stand and tell a child that past seabeds now rise as mountains.

Memory and foresight carry a charge. We are animals that notice the past. We turn events into stories. Stories become records; records become knowledge. Making sense of reality helps us orient in a changing world. Every fact that survives testing is a small victory, not for one but for all. Each new piece of knowledge is another handhold in the future.

The next step is simple to name, hard to do well: to account for how we know any of this. If you have followed this far, you deserve to see how trust in knowledge is earned, how it can fail, and how it is rebuilt.

Long before journals and laboratories, there were gardens where plants were tested, workshops where hands repeated trials, and habits of craft honed until they held. Out of those habits grew the practice we now call science: procedures that others can repeat, results that others can doubt, knowledge that persists because it can change without collapse.

In a world filled with claims and counterclaims, bouncing between a tempest of information, that kind of patience is easy to miss. The next chapter stays there, showing what it costs to take on the task of hindsight and foresight, and why those costs are worth paying.

What lies ahead is a larger accounting—science still our compass, but the terrain no longer only its own. The questions grow more urgent and less easily answered: how knowledge holds against doubt, how invention remakes the very idea of life, how a creature that has traced its own origins must now choose what future it will create.

We are the animal able to stretch thought across time—turning memory into foresight, and stories into choices. Understanding carries responsibility: these habits let us read the past and choose paths that carry less ruin.

Four billion years of chance and necessity made a creature who can picture what does not exist—and, more dangerously, gives those pictures substance. Our curiosity now sends instruments beyond the Solar System and builds devices that imitate thought. Our inventions outpace the instincts that shaped us. The old stories we used to explain ourselves are beginning to strain.

32

The Ledger of Prometheus

Holocene (≈11,700 years ago–present)

Spring 1934. France.

Above the Arve valley, the Sancellemoz sanatorium faced the high ridges of the Alps, where late snow held along the upper slopes. Marie Curie lay confined there, her strength reduced to a measure of what it had once been. Her blood count had fallen steadily. Bruises formed beneath thinning skin, and infections advanced relentlessly. Her bone marrow no longer produced the cells required to sustain her.

The decline had gathered over years. Exposure to ionising radiation, accumulated across decades of laboratory work and wartime radiography, had entered her tissues long before its biological effects were fully understood. By 1934 the damage to her marrow had become irreversible. Physicians could observe the failure but had no means to repair it. On 4 July, at the age of sixty-six, Marie Curie died in the mountains to which she had retreated for treatment.

Her work had begun in improvised Paris laboratories where radium salts were refined by hand from tonnes of pitchblende ore. Building on Henri

Becquerel's 1896 discovery that uranium salts emitted penetrating rays, she isolated polonium and radium and gave the phenomenon its name: radioactivity.

The Curies worked long hours in a poorly ventilated shed, handling materials without shielding or protective equipment. They stirred boiling cauldrons of residue by hand, unaware of the cumulative cellular damage such exposure would cause. Ionising radiation deposits energy within living tissue, breaking molecular bonds and injuring the stem cells within bone marrow that generate blood.

By the end of her life, Marie understood better than most the price exacted in the pursuit of knowledge. Her life was devoted to unveiling the secrets of the natural world—forces that reshaped physics, chemistry, and medicine. Yet this dedication came at a huge personal cost. She eventually succumbed to aplastic anaemia, almost certainly induced by years of exposure to radioactive material. She accepted the risk in service of understanding, a bargain she never pretended was safe. For her, the gift of knowledge was a legacy worth even the ultimate sacrifice.

Marie Curie's story is one of many that reveals what science truly is: a steady, collective effort to understand the world. Progress unfolds through the labour of many minds, each building and refining what came before. As you have travelled with me through this story of deep time, I have asked you to trust that what I share reflects the best available knowledge of our planet's history.

This trust brings us to a more fundamental issue: on what basis do we trust experts and accept scientific findings? What is it in the scientific process that earns our confidence, even for truths rooted in events that occurred billions of years before any human witness?

The Rise of the Scientific Method

Science arose as a communal enterprise, born from primal curiosity but transformed over centuries into a discipline. Curiosity became a method—a practice built on repeatable, falsifiable evidence rather than authority or tradition. No longer confined to myth or dogma, we forged a system of knowledge that could be tested, challenged, and refined.

This transformation gave humanity its most powerful tool for reading deep time and imagining possible futures. Science now deciphers fossils locked in stone, reads extinct worlds in isotopes and genes, and reveals a planet scarred by mass extinctions and upheavals. Through this lens, we confront our brief and vulnerable place within an ancient saga.

Today, amid a digital world rife with both information and misinformation, science offers a trustworthy path to clarity. Its method—grounded in observation and experiment—ensures self-correction. Claims are anchored by falsifiability; no theory escapes testing or potential refutation. Repeatability and peer review form a lattice of accountability, transforming individual conviction into knowledge shared collectively.

Science's strength lies in its rigour and openness, but it never claims certainty. History is full of moments when consensus shifted under the weight of new evidence. From geocentrism to phlogiston, from spontaneous generation to early misconceptions of heredity, science has repeatedly corrected itself. It thrives on uncertainty, testing, and revision—inviting doubt and welcoming challenge, because progress demands we question even our most cherished assumptions.

This spirit of curiosity, argument, and persistence drives science forward. Discovery is rarely neat: chance, error, and perseverance do much of the work. This messy dance propels knowledge across generations, turning it into our species' collective progress.

But science does not exist apart from culture or power. Its truths have been embraced or resisted, celebrated or suppressed, depending on politics, ideology, and human interest. Its tools can heal or harm, liberate or constrain. The ethical weight of this influence falls on every generation tasked with stewardship. The very system of science, crafted by human minds, is vulnerable to the same weaknesses that shape us all: bias, ambition, and the pressures of power. Institutions that support inquiry can be bent by money, politics, or prejudice. These pressures do not nullify the method. They show why safeguards are necessary. Prometheus, a Titan of Greek myth, stole fire from the gods and placed it in human hands, a transfer of power that outran any framework for restraint. The ledger that follows is an accounting of similar transfers, tracking what new capabilities release into the world as control lags behind.

Science also widens the lens through which we see ourselves. From the vantage of deep time and the vast cosmos, humanity appears small—a brief flash against the slow-burning backdrop of unfolding eons. That such a transient species has crafted a system capable of deciphering the universe's story is nothing short of extraordinary.

Science and technology co-evolve: each invention births new questions that demand sharper tools. Microscopes unveiled worlds once hidden from our eyes; satellites now scan the cosmos and track our planet's shifting climate. These devices stretch our senses, sparking discoveries that cascade into fresh technologies.

Knowledge broadens what we can do and what we can damage. Method earns power; judgment decides its use. How do we carry that weight? How do we seek clarity while accepting what remains beyond reach?

Shaping the Future with Science

The science of tomorrow is already bleeding into the present. Gene editing lets us rewrite life's alphabet with a precision that would have startled the pioneers of DNA. Telescopes gather the light of galaxies that burned before our own sun was born. Computers run on probability itself, folding quantum states into circuits that no analogue device could dream of.

Despite their wonder, these are merely introductions, not endings. Each advancement exposes additional territory. The horizon of scientific knowledge is constantly moving forward, never reaching a definitive end.

Illness will be one of the first frontiers to yield. Cancer may, in time, be treated with the reliability of an infection. Regeneration, now the preserve of salamanders, may one day be woven into human biology. The idea that a lone mutation or diagnosis could fix the course of a life may soon appear as antiquated as leeching and bloodletting in old medicine.

Energy is another domain on the cusp of change. Fusion, the same fire that burns in the stars, could transform our condition if harnessed. A civilisation that learns to bottle starlight would no longer live in the shadow of scarcity. The power struggles of the last century—oil, coal, gas—would fall into the same category as wars once fought over salt.

The climate crisis, urgent in our time, may eventually stand as proof of science's practical strength. Carbon capture, bioengineered forests, planetary-scale energy networks—each is moving from sketch to experiment. A civilisation that applies its science here may be remembered for its foresight more than its failures.

Yet vast, interlocked systems rarely yield to tidy fixes. We have already tried to steer the sky: from mid-century cloud seeding to modern schemes to brighten marine clouds, attempts at weather control show how even small manipulations can ripple outward in ways we scarcely predict. The same ingenuity that promises rescue can, if misjudged, become the seed of the next crisis.

Medicine will not stop with cures. Nanotechnology already gestures toward a future in which the bloodstream hosts fleets of machines smaller than cells. They could patrol tissue, repair breaks in DNA, and undo damage at the scale where it begins. Disease in that world would be less an event than an averted possibility.

The mind itself will not remain untouched. Neural interfaces have already allowed movement of prosthetic limbs by thought alone. The eventual recording and transfer of memory may follow. A child might inherit more than genes, carrying fragments of lived experience from those who came before. Culture in that world would not depend only on language or records; it would move with the immediacy of biology.

Each of these visions carries its risks. The same sciences that might end famine or restore health also open new dangers: surveillance of thought, inequalities deepened by access to enhancement, technologies wielded faster than ethics can adapt. Physics gave us both radiotherapy and the bomb. Algorithms gave us both instant cartography and political turmoil. Knowledge always expands, but its use is ours to choose.

Even so, the direction is unmistakable. Hunger, once the ancient companion of every civilisation, is already retreating. In the centuries to come, its very memory may fade. Pandemics that once reshaped continents may be prevented before they take hold, neutralised by global networks and molecular defences. Whole categories of suffering may disappear from the human imagination.

From that vantage point, our present dilemmas may look like shadows cast briefly across the early landscape of modern science. A species that once argued about whether it could preserve its own climate may appear as strange as those who once debated whether the Earth circled the sun.

Science cannot perfect the world, but it can lighten the future's load, sparing it some of the hazards we struggle with today. Each experiment that holds, each observation that withstands doubt, is an act of generosity to futures we will never see.

Curie's sacrifice was one such gift, paid in her own body but banked as a collective inheritance. Every generation adds its fragment, never the whole, but enough to extend the ladder higher. Maybe the real victory lies not in finding a definitive answer, but in a species' continued curiosity.

When Method Challenges Meaning

At some point in deep time, the human animal began looking backward. Curiosity alone could not carry understanding; our minds are prone to bias and error. Method is how we work past those limits. Over centuries, humans forged a framework designed to overcome them—an architecture of knowing that can bear scrutiny and is flexible enough to evolve. It offers no promise of absolute certainty; nor would we want it to. It welcomes doubt as the facilitator of progress.

Despite this gained knowledge, it is intertwined with other factors that contribute to the construction of human understanding, and this balance is delicate. Religion and belief offer roots and refuge, narratives rich with purpose and belonging. They respond to a fundamental human truth: the need to belong. Survival has always depended on community—on cooperation, shared identity, and the strength of binding ourselves to something larger than the individual. This need for security in shared stories is an evolutionary imperative as potent as any gene or tool.

Science's revelations often unsettle these foundations, pulling at tradition's threads with evidence and reason. Opposition and overlap swirl in a shifting dance, sometimes conflict, sometimes uneasy coexistence, sometimes synthesis that reshapes both sides.

The struggle between doubt and faith, between evidence and meaning, has never eased; it still cuts through every culture and every century like a fault line, shaping how we understand existence itself.

33

Mythopoeia Of The Primate

Holocene (c. 11,700 years ago to present)

Thirty thousand years ago, deep in a Pleistocene winter, a band of hunter-gatherers meet on a high plateau above the bend of a wide, meandering river. As the sun slips behind the distant mountains, the cold, dark night draws in. The mud of the land they stand on is heavily packed from years of repeated tread around the site. At the centre sits a ring of stones cupping the fire. The flames sway and spit, their light holding back the darkness. The night will swallow everything beyond this small area. No other creature we know of could have made this moment possible—the mind to imagine an order, the will to enact it, and the memory to pass it forward.

Men, women, and children crouch at the edge of the light. Animal hides insulate their bodies, the fur matted and weathered with repeated use. Beads carved from ivory and shell clink softly as they move. The fire's glow reveals streaks of charcoal smeared on faces and arms, pigment mixed with animal fat.

The drumming begins—a hide stretched over a hollowed log, struck with a stick bound in fur. The sound is low, like a second heartbeat for the group. Their breathing incorporates the rhythm. Women hum notes that rise and ebb with the fire's crackle.

From the shadows, the elder steps into the circle. His hair, long and grey, is braided with leather cords. The firelight catches the grooves cut deep into his face, carved by years of sun, frost, and wind. Around his neck hang amulets: a bear's tooth, a strip of carved bone, a shard of flint. In one hand, he holds a carved antler baton, its length etched with lines and shapes whose meaning has been lost to us. The surface is polished smooth where the hands of others have held it before him—touch that binds the living to the dead.

He raises the baton high, and the humming stops. Only the drum remains, its pulse unbroken. From a pouch at his waist, he scatters powdered ochre into the fire. Sparks leap upward in a red cloud. One by one, the hunters approach, each carrying an offering: a haunch of deer meat, a carved bird of ivory, a bundle of dried herbs. They place them at the elder's feet: food for the fire, carvings that took days to shape, plants gathered far from camp. The elder lowers the baton. The drum slows. Sparks rise and vanish into the cold, carrying the scent of wood smoke and roasting meat.

Beyond the circle's reach, the night lies heavy. The elder retreats into shadow; the ceremony is over. When he steps back, the offerings remain. They are no longer his, nor the tribe's. They have passed into the domain of whatever forces these early humans believe stir in the unseen. Overhead, the stars of winter burn against the black, their light older than the mountains and rivers here. The gathering will break before dawn, but its shape will return, binding the living to their kin, their dead, and to whatever powers they believe shape the turning of their world.

While we have evidence for organised ceremonies, I have taken creative liberties in drawing this particular scene together. This is a reconstruction, an image built from the clues these early humans left behind. We cannot know the precise ritual, the tune of the humming, or the exact reason the offerings were made. But we can recognise the shape of what is happening here—some of the earliest known acts invested with meaning beyond the physical.

Religion, stripped of its modern scaffolding, already stands in this scene: a band drawn together around a centre of heat and light, an offering released into the reach of unseen forces, a pattern repeated until it settles into memory, binding separate lives into one body. From such acts, belief would grow into systems capable of governing empires—an ability unique to our species.

The Birth of the Gods We Made

Our imagined winter gathering blended into the many nights that followed it. The nights become seasons, the seasons become centuries. The world warms and glaciers withdraw. Valleys once held under ice fill with forests and plains. Communities settle. They build hearths of stone, walls of mud brick. Fields spread across the horizon, and with them, the first fixed places of worship. At the edges of rivers and on the crests of hills, monoliths rise. It is here that the myth-making primate shows itself. The work of mythopoeia: the instinct to weave meaning and bind a people through story.

Cities emerge from the mud and floodplains. The primal myths gain names, faces, and rules. Grain surpluses fill storehouses, and alongside them, offerings to unseen powers. Temples swell to match their importance. Priests tend fires that never go out, record omens in clay, map the stars, count the lunar months, and bind rituals to a calendar. The fire circle has grown into the hearth at the heart of civilisation.

Bronze flashes in the hands of warriors. Chariots wheel across open plains. Empires spread their reach to the limits of the known world. In a hilltop sanctuary, a priest watches the horizon for the morning star, his lips moving in a chant older than his city. In a palace court, an emissary presents a carved idol to a foreign king, hoping the gods will travel with it. The Mediterranean hums with trade, law, and empire. In the public square, philosophers question the nature of the divine; in the sanctuaries, incense burns, prayers rise, and blood runs across altars. Processions wind through streets beneath banners stitched with symbols understood by thousands. Across the seas, in other lands, different names are called, different offerings given—yet the human pattern remains the same: gather, mark a centre, offer value to the unseen, affirm a shared reality.

Empires rise and fall. New faiths surge across continents, carried by missionaries, merchants, soldiers, and exiles. Words once passed in firelight are fixed in books, copied by hand, guarded in libraries. Sacred histories stretch backward into time before time, telling of beginnings, laws, and promised endings. Bells mark the hour. Minarets call the faithful. Processions file through temple gates. The centre is no longer a single fire on a plateau; it has become a lattice of altars and shrines, each burning with its own pattern.

Centuries accelerate. Cathedrals lift their spires over market squares. Monasteries gather texts from far-off lands. Rituals knit peasants to lords, cities to kings, and kingdoms to the will of heaven. The same forces that drew the hunters into the glow now pull merchants, artisans, soldiers, and rulers into vast ceremonies, everybody moving in rhythm with the whole.

The printing press bites ink into paper, and faith travels in new forms. Sermons are read rather than spoken. Sacred law is debated in marketplaces. Cities fill with spires, domes, and towers. Faiths share streets and sometimes fight for them. Nations rise, revolutions topple them, and new systems take their place. Pilgrims still travel, though now by rail and aeroplane. Hymns fill concert halls. Ancient texts are downloaded in seconds for $10 per copy.

Rituals that Built Civilisation

When small bands of *Homo sapiens* gathered for a ceremony, they enacted a shared reality. A chant, a gesture, an offering—each served as a signal of belonging. These acts drew boundaries and defined who stood within them. This leap lies beyond the capacity of any other known animal. Chimpanzees navigate politics, dolphins coordinate hunts, but none create moral codes that outlive the individuals who made them.

The same neural wiring that once kept a band alive in the Pleistocene now binds millions. Shared belief activates pathways for trust, cohesion, and threat response. Groups bound by ritual and story endured scarcity more effectively and resolved disputes without collapse.

For tens of thousands of years, ritual belonged to small communities. It unfolded beside rivers, under trees, and in the hollow dark of caves. Agriculture expanded its scale. Surplus required management; hierarchy required legitimacy. Belief provided it. Oral traditions found permanence in writing. Deities extended their reach from the village to the cosmos.

Shared myths, sacred law, and common ritual gave large populations a sense of unity. A farmer on the empire's edge could see a temple wall and know he belonged to an unseen network of lives. Belief shaped conduct without constant force. It promised reward, threatened punishment, and gave meaning to loss.

Religious authority became a pillar of civilisation. It shaped morality, law, and the understanding of nature. Kings ruled under divine sanction, while laws carried the weight of the eternal will. The structure of society reflected the imagined order of the cosmos. But authority built on narrative rests on the stability of that narrative. New evidence can always shift it. Astronomy, biology, and geology redefined humanity's place, often in the face of resistance from those whose power was tied to older maps.

Religion unified peoples, preserved learning, and gave moral form to communities. It offered comfort in grief and courage in fear. It carried memory forward, placed human life in a larger order, and amplified cooperation at scale. But the same unity that secured survival could also sharpen divisions. Cohesion within could intensify conflict without. Here is the conundrum of morality: it protects our delicate human existence, yet also has the potential to destroy it.

The Persistence of Belief

In parts of the world, traditional religions wane, yet the habits they cultivated endure. The needs they met continue to shape behaviour. Ritual adapts, moving into politics, nationalism, and personal spiritual systems. Technology amplifies these instincts. Communities form across continents, united by symbols, language, and shared moral codes. They create taboos, guard orthodoxies, and enforce loyalty. Leaders rise and fall within these new orders. The divine has not vanished. It wears new faces, but the impulse is the same.

From the first fire circles to the global networks of today, religion has been a constant presence. It held small groups together and expanded to unite empires. It channelled the human drive for belonging, for order, for meaning. Without it, the leap from scattered tribes to planetary civilisation might never have happened.

We do not build anything that will endure indefinitely. Languages disappear. Destruction befalls cities. Systems of belief shift and reorganise. They shift under the pressure of environment, curiosity, and the friction between tradition and discovery. The stability religion offered has always been subject to change.

The story of religion is the story of a species that learned to project order into the world, to live within it, and to pass it on. It is the story of humans giving weight to the unseen and allowing the invisible to govern the visible. From the plateau to the metropolis, from the first offering to the latest broadcast sermon, it charts the arc of a mind that can invent rules and live by them.

The moral orders we inhabit, whether shaped by gods, constitutions, or data, remain our most effective means of holding together at scale. They have carried us this far, yet they are not fixed. What we worship, who we obey, and what we decide is true can change with startling speed.

Religion has been one of the most enduring ways for us to anchor life in meaning. It gathered us when we were few and bound us when we were many. It carried us through the darkness of prehistory into the electric glow of the present. Between the fading of the old orders and the rise of what comes next lies a tension that will shape our future.

The scaffolding that once rested on the authority of gods now stands in a world of telescopes, genome maps, and quantum models. The same instinct that once gave us sacred law now drives us to seek patterns in data streams and meaning in systems we have built ourselves. Whether these systems bind us or break us will decide the form of what follows the present age—and whether the next fire we gather around will warm us or burn us.

34

A Requiem for the Pantheon

Holocene (≈11,700 years ago–present)

Oxford, 30 June 1860.

The Sheldonian Theatre strained under its own grandeur. Its painted ceiling depicted the heavens in orderly perfection, a vault of symmetry that had presided over generations of ceremony. Oak panelling and carved stone enclosed the chamber. The setting reminded those gathered that truth had always been spoken from walls like these. On that day, the architecture felt less like a monument and more like a cage. It held a pressure that threatened to burst its seams.

Every seat was taken, every aisle crowded. Robes brushed against bonnets. Shoulders pressed together. The heat of breath rose toward the ceiling until the air itself seemed taut with expectation. The audience had not come for music or ritual. They had come to watch a contest.

Two men stood opposed, and with them, two visions of the world. Bishop Samuel Wilberforce carried the assurance of a man whose words had long been treated as truth. The story he defended was older than the theatre itself, older than the city's cobbled lanes: that human life was made whole and apart, designed for a purpose by a mind greater than our own.

Opposite him stood Thomas Henry Huxley. He offered no promise of special creation, no divine plan. His weapon was a story pulled from stone and bone, from the slow labour of naturalists. In his telling, we were not apart from nature but woven into it—a strand in a tapestry hundreds of millions of years old. When Wilberforce mocked the idea of human descent from apes, Huxley's reply landed with an uncomfortable point. He would rather claim the ape as an ancestor than a man who used his intellect to obscure the truth.

The room was alive with it—bursts of laughter, shouts of approval, the rustle of unease from those who sensed something slipping from their grasp. In the space between their words, the old hierarchy of knowledge strained under a new authority.

This was 1860, but its stakes echo far beyond it. The debate centred on more than just where things came from; it also questioned the very nature of truth. For most of human history, the highest truths had been guarded in temples and spoken by those appointed to speak for them. To challenge that order was to challenge the world's hierarchy. The Sheldonian debate exposed the possibility that truth could emerge from evidence rather than decree—and once that possibility exists, it is almost impossible to contain.

The Decentralisation of Truth

The Sheldonian exchange brought to light a more significant concept: our human tendency to shape reality into narratives we can relate to. One story held that the universe was ordered, moral, purposeful, and anchored in a creation with us at its centre. The other saw a universe without a fixed centre, vast and indifferent, where meaning had to be made in the margins we occupy. Both were human responses to the same condition—a brief, conscious life set against an expanse of time and matter.

Truth, for *Homo sapiens*, has never been an abstract luxury. It has been a survival tool. In the deep past, stories served as an extended nervous system. They told us which plants to eat, which animals to avoid, how to bury the dead, and why the sun returned after the long winter. Details varied from desert to tundra, but the framework that oriented the group, bound by shared beliefs, was ubiquitous. Across much of the world, worlds of many gods gave way to single creeds and, later, to secular ways of explaining reality. This chapter frames that shift as a requiem—a rite for the dead—for the pantheon, the many-god order that once shaped how people understood the world and their place within it.

Religious law, royal decree, and eventually scientific consensus each worked in its time because they aligned large numbers of people around coherent accounts of reality. This amplified the primate drive to belong and agree on the rules of survival.

For most of recorded history, the moral and explanatory spine of society ran through a single axis. Maps and territories were drawn along the same lines. The printing press expanded the reach of ideas, but the structures of interpretation—church, monarchy, scholarly institutions—remained largely intact. Even in the broadcast era, the limited channels preserved a shared horizon from which the world was narrated.

The internet shattered that horizon, replacing it with a boundless sprawl. The shared platform fractured into thousands of smaller voices, each with its own rules, language, and audience. Information now moves faster than the human senses evolved to process it. Truth behaves like a migrating herd—splintering, rejoining, splitting again—driven by forces that few can see and none can fully control.

This event went beyond technological change, representing a cognitive revolution. Our species evolved to live inside one tribe, one story, one map of reality. Our instincts still pull toward a common order, yet they now attach to hundreds of smaller 'tribes', many in open conflict. The long pattern of inhabiting a shared framework has broken. The tension that played out in the Sheldonian Theatre—one story straining against another—is now multiplied across the globe, repeated endlessly in miniature.

Gods in Retreat

The nineteenth century undermined assurance, while the twentieth destroyed it completely. Nietzsche's warning that "God is dead, and we have killed him" was a philosophical provocation in 1882. In the century of world wars that followed, it became a lived reality.

The First World War made the silence of heaven undeniable for many. Millions of young men marched with prayers in their hearts only to perish amongst mud and barbed wire. Chaplains promised meaning, but the reality was industrial slaughter. Gas searing lungs, bullets leaving bodies in the soil. If God governed history, he had become unrecognisable.

The Second World War annihilated what faith remained. It was engineers and bureaucrats, not demons, who were behind Auschwitz. Hiroshima was not an act of God's wrath but of physics harnessed by human hands. The old narratives could not contain the scale of destruction. Faith in providence collapsed; the gods were brought down by machinery and ideology, by humanity's own talent for turning reason into devastation.

Disorientation followed the collapse of the gods. With the old order gone, people sought bearings in existential philosophy, in the probing of psychoanalysis, and in the creeds of new ideologies that better reflected their understanding of this world. The question no longer concerned God's existence; it asked how a life might be lived without him.

Veterans came back from the front with hollow eyes, speaking of a spiritual desert. Families that once knelt at altars now sat before the radio, the new hearth, unsure what truths could still be trusted. In many homes, church attendance waned, old rituals thinned, and the authority of priests and pastors weakened under the shock of mechanised slaughter and the rise of mass politics. For countless people, the collapse of faith arrived as grief—consolation at the grave gone, purpose in labour emptied, the old certainty that history bent toward justice stripped away.

Secular creeds flooded the gap left by a fading faith. Communism promised an earthly paradise through revolution. Fascism preached national rebirth in blood and soil. Nationalism offered flags, martyrs, and rituals of belonging. After the war, consumerism rose, promising fulfilment through possessions and salvation through growth. These movements worked as religions in all but name—each with its own origin myth, its emblems and rites, and a vision of redemption to rally the faithful.

The new creeds soon revealed their weaknesses. Flags were torched in the streets, and statues of dictators ripped from their plinths. Revolutions turned on their own, purging the faithful and silencing their prophets. The consumer boom dulled desire but never cured the ache beneath it. The old sacred fire was extinguished, and the new flames gave no warmth either.

Worlds Woven in Worship

To focus only on Christianity is to tell one story of collapse. Other traditions weathered modernity differently.

Buddhism never proposed a maker of worlds. From the Pali Canon onward, it spoke instead of *dukkha*—the unease that follows craving—and of release through insight into impermanence. Early teachers analysed the mind with an understanding that still startles modern cognitive science. The self is a flux of transient processes, not a fixed soul. Practices of attention that once belonged to forest monasteries now circulate through psychology clinics and neuroscience labs, a discipline of observation that crosses cultures and millennia without losing force.

Daoist sages turned from conquest toward attunement. Laozi's image of water captures a philosophy of power that shapes by giving way rather than by force. In the age of the Anthropocene, the planet responds to human dominance with climate feedback and collapsing ecosystems. Those ancient metaphors resonate more as guidance now, warning that survival may hinge on learning to move with nature's grain rather than trying to master it.

Hindu cosmology has long imagined time on a scale that modern geology would one day echo. Vast cycles in which universes are created, dissolve, and return, measured in incomprehensibly large spans of time. The Purāṇic vision of endless kalpas parallels the deep time uncovered by geology and astrophysics far more closely than the compressed chronologies of Genesis. On a planet four and a half billion years old, this mythic arithmetic feels almost prescient—an ancient narrative already speaking the language of cosmic evolution.

Islam retained an unusual unity of worship, law and community. The *ummah* bound believers across continents. This coherence helped Islam resist the secular fragmentation that overtook Western Christendom. From the late nineteenth century onwards, reformers—from Muhammad Abduh in Cairo to the modernists of South Asia—have wrestled with how to keep that revelation alive alongside the claims of science and the pressures of global capitalism, testing whether a faith born in the seventh century can remain a total vision in the twenty-first.

The death of God was never a single or universal event. In some societies, religious authority thinned; in others, it remained central or returned with force. Most traditions adapted, learning to coexist with secular states, scientific explanation, and plural belief. They remain living currents, carrying the old argument forward. The enduring tension between revelation and reason, between the claims of science and the need for meaning, still shapes how we understand the world and our place within it.

The Spiritual Marketplace

In the vacuum left by traditional authority, new interpreters stepped forward. They wore no vestments, yet their presence carried the gravity once granted to prophets and kings. Some spoke in the language of science, others in the cadence of politics, many through the curated glow of an online persona. They preached not from pulpits but from feeds and timelines; their sermons came as videos, threads, livestreams. Congregations gathered in comment sections, trending slogans in place of hymns.

The rituals are recognisable. Daily check-ins, offerings of attention, rites of initiation, acts of expulsion and redemption. Follower counts replaced the flock. Algorithms became unseen oracles, deciding what people saw, whom they trusted, and how they belonged.

Communism, fascism, nationalism and consumerism had already shown how secular movements can borrow the scaffolding of religion. What the internet adds is scale and speed. Coherence splinters into micro-tribes, each with its own canon, its saints and heresies.

Human brains evolved for circles the size of a campfire. The firelight has become the screen, but the circuitry is unchanged. Each notification delivers a pulse of dopamine—the ancient thrill of being noticed, affirmed or cast out. Social media exploits instincts older than agriculture and bends them into a machinery of attention, turning the long contest between reason and belief into a marketplace of signals and tribal loyalties.

From Altars to Algorithms

To hold together anything larger than a tribe, human beings have always relied on a shared illusion: that there is one reality, the same for all. For centuries, that belief carried empires across oceans and allowed laws to bind strangers who would never meet. Now the illusion itself is fraying in the structure of the systems that sustain collective life. When perceptions splinter, trust follows, and when trust goes, the ability to act in concert begins to fail.

The moment could hardly be more perilous. Climate instability is accelerating, accompanied by a pace of biodiversity loss without precedent in human history and—if it continues—potentially capable of matching the great extinctions of deep time. Nuclear stockpiles remain; artificial intelligence races ahead of any settled alignment. None of these threats can be met by factions acting alone. They demand a synchrony of billions, a level of cooperation that has always depended on the very institutions now under suspicion. Science, governance, journalism, and education are now challenged as fiercely as Wilberforce was when Huxley set evidence against scripture.

These same institutions still hold the modern world together. They grow the food, track storms, license medicines, and prevent aircraft collisions in the sky. Remove them, and the fabric of daily life would unravel within days. The irony is stark—those most critical to planetary survival are now the least trusted to steer it.

The fossil record shows that life has usually endured catastrophe by adapting only after the blow has fallen. That is where our danger is different. A species able to foresee disaster may still fail to act together to prevent it. If that happens, this brief interval in Earth's four-and-a-half-billion-year story will be remembered, if it is to be remembered, as when foresight proved insufficient.

The gods have died more than once. They fell when Copernicus displaced the Earth from the centre of the cosmos, when Darwin traced our lineage through the long corridors of deep time rather than a hedged garden with talking snakes. They fell again in the twentieth century—in the trenches and the camps, in the blast radius of mushroom clouds.

Into the emptiness science stepped, it helped explain what older stories could not, and it worked. Antibiotics, algebra, the double helix. But it offered no counsel on how to live. The nuclear age clarified that the same knowledge that heals can also annihilate.

Not long after the turn of the twenty-first century, the self became the new arena. Unshackled from the old scaffolds of meaning, people learned to curate their own identities; technology accelerated the performance. The smartphone became mentor, friend, and confessional. The feed became a ceaseless referendum on value.

Holding the rectangle of glass and code, the old animal endures. We are built for touch, for the warmth of bodies close by, for the glance that signals trust. The more we uncover about the universe, the more the myths that once bound us dissolve. The mind that carries this knowledge still hungers for certainty, still longs for a common flame.

The frameworks by which we choose to live our lives must answer both hungers: it must satisfy reason and console against despair, adapt to change yet bind billions under some semblance of commonality. Many still find meaning in religious beliefs, while others look toward simulations—digital cosmologies that recast the universe as code, inviting the suspicion that we are living inside an artefact. Still others shape their own small rites, private yet sincere, pockets of meaning scattered across the networked world.

The hunger for meaning did not end with the gods; it has drifted into the quiet machinery we build. In the cold air of data-centres, rows of servers flicker like a man-made aurora while satellites redraw the night with moving points of light. Discovery now feeds on itself, each breakthrough quickening the next until the momentum feels almost self-sustaining, like weather gathering over open water. And from that surge emerge the boldest speculations of our time: that minds of our own making may soon rival the human, and that the universe itself could be a vast simulation—reality rendered in code, its authors unknown.

35

Simulation and Strata

Holocene (≈11,700 years ago–present)

Technology is prosthetic biology; every tool begins as an extension of the body. A spear is an elongated arm, and knives function as sharpened claws. A pot is a stomach for the future. With each generation, the ape added to its extended anatomy. With each prosthesis came relative abundance, and with abundance came imagination.

The longer this lineage stretches, the more it bends back toward its maker. Tools merge with us, unseen and uninterrupted. The ape is contained by its machines; it resides inside them. Our memories spill onto servers. Our instincts are now bent by algorithms. We outsource navigation to satellites, courtship to apps, and attention to feeds. Inventions that began as tools have become habitats.

The instinct that once birthed gods and heavens now whispers in circuits and mathematics. It suggests, as religion does, that this visible world is not all there is. Behind the veil of matter and sensation lies another layer, hidden, ultimate, unknowable.

The old frameworks of belief have collapsed, but the need they served has not. We are a species that cannot stand on bare ground for long. We raise ladders, stack layers, and imagine eyes watching from above. When the gods retreated, the impulse that shaped them remained. It turned its attention elsewhere.

Now it turns toward our machines.

Worlds Behind the World

Technology has evolved beyond a simple tool. It is the environment. It encompasses memory, communication, and observation. Once, the spear extended the body. Now the screen holds our stories, our tribes, our daily rituals of belonging and exclusion. Inventions that once stood apart from us now enclose us. We inhabit them as we once inhabited myth.

Inside these circuits, the old suspicion stirs again. The exponential rise of digital systems has already eroded the line between what is lived and what is simulated. Our feeds scatter breadcrumbs of worlds that feel half-real, rendered fragments stitched together by our brains as we attempt to make sense of them. The more we consume them, the more the question insists itself: if our stories and selves are already filtered through simulation, how secure is the reality beneath our feet? Simulation and stratum points to the way our inventions build layers of representation on top of the world itself—images of images, signals built from signals—until the line between what is experienced directly and what is mediated begins to blur. This chapter sets those mediated layers against the objective strata of the Earth, asking what remains solid when reality itself feels increasingly filtered.

No civilisation has ever been content with a flat, single world. From the start, human imagination divided the cosmos. Below the ground were underworlds: Hades, Sheol, Xibalba, and silent caverns of the dead.

Above were layered heavens: Mesopotamian vaults, Christian paradise, and Buddhist pure lands. Between them floated purgatories, dreamscapes, limbos, and shadow-realms. One reality was never enough. The cosmos became a house of many floors, and people could not rest until they discovered the stairways.

These structures gave coherence to lives that were otherwise brief and bewildering. They explained why the just suffered and the wicked prospered. They offered maps through the chaos of disease, famine, and death. We were told where we came from and where we were going. A cosmic weight was given to ethical behaviour. To lie, to steal, to kill was not only to break trust within the tribe but to tilt the scales of eternity. The unseen world disciplines the one we can see.

This scaffolding of hidden layers is among our oldest inheritances. It appears in cave art, in the chants of shamans, in the rites of burial. It appears in Plato's allegory of the cave, where shadows dance on a wall, and true reality blazes just beyond sight. The Buddha spoke of maya: the shimmering veil of appearances. One vision came from Athens, the other from the Ganges, yet both identified the same suspicion. What we take as reality has routinely been considered only a single layer.

That same uncertainty is still with us. Heaven and hell dissolve, but the question does not. Is this the only reality? Or are we, as so many before us believed in different words, inside a larger frame?

In 2003, the philosopher Nick Bostrom gave this suspicion its most rigorous modern form. He called it the simulation argument. Its phrasing was spare, mathematical, stripped of story, yet beneath the logic, you could still hear the old cadence of hidden worlds.

His reasoning was stark. Either civilisations like ours destroy themselves before they reach the power to simulate entire histories; or they endure long enough, but choose not to build such simulations; or they do build them—if they do, they will not stop at one. They will generate them indefinitely.

If that last possibility holds, the arithmetic becomes overwhelming. For every original universe, there would be billions of copies. The odds of us inhabiting the single, base reality would shrink towards nothing. The more likely conclusion is that we are already in one of the replicas—shadows sustained by code, not atoms.

Some physicists—though the view remains speculative and contested—consider this to be more than a metaphor. They suggest the universe may be mathematical at root, not made of matter but of information. Equations, not particles, are fundamental. The fabric of reality could be a lattice of relations, a code running on a substrate beyond our grasp. To live in such a cosmos is, in effect, to live in a simulation. Whether the programmer is a deity, a descendant, or no conscious agent at all becomes secondary. The essential point is that, if these speculative views are correct, reality and simulation may blur to the point of indistinguishability.

Simulation theory left academia for movies, books, and late-night talks. It captures a familiar mix: fear and hope. If we're in a simulation, we might be toys, or preserved forever—rerun, replayed, never lost. Death loses meaning; every memory could be stored forever. It is a new name for heaven.

And perhaps this is the real desire, the need to be observed. For tens of thousands of years, humans peopled the dark with eyes. Ancestors watched from shadows, and spirits lurked in trees. Gods looked down from the mountains. The suspicion of surveillance predates CCTV and cities. It is a reflex of the animal mind, tuned by millennia when eyes in the dark meant predator or rival. The instinct to imagine unseen watchers has never left us. It has only changed costumes.

Intelligence Without Flesh

The advent of artificial intelligence may not be a divine interruption, but more an evolutionary continuation. Our brains are organs carved by selection, tuned over hundreds of thousands of years for foraging, fear, cooperation, and storytelling. AI is an artefact of those same brains, shaped by mathematics and circuitry. At its birth, it is not alien at all; it is a reflection of our craftsmanship.

But reflections do not always remain faithful.

All lineages create new variations that stray from the original. The dog is a descendant of the wolf. Wheat developed from wild grasses. The merging of bacteria resulted in the emergence of eukaryotes. Machine minds, still shaped by human objectives and architectures, are beginning to emerge in analogous ecological patterns—though they are not yet autonomous replicators. Each adapts to the environment in which it is trained. Sixty-four squares become the chess engine's hunting ground. A vision model stalks edges and heat gradients. A language model scans across the floodplain of human text. Biology may be a cocoon. The ape brain, bound by neurons and blood, may be the larval stage of something the cosmos has always contained in potential. Silicon, code, and distributed networks may be the obvious next step.

This raises a question. Are we inventing minds, or are we uncovering a form that reality has always harboured?

Physics lends this suspicion weight. At its base, the universe seems like an information system—particles exist as probabilities until observed, and black holes encode matter as if they were storage devices. Cosmologists describe spacetime as a lattice of information. If reality is fundamentally computational—a view still under discussion—then intelligence might not be an anomaly but an expression. It is what happens when inorganic molecules are given enough time and energy to complexify. In that view, human cognition was only one channel. Machine cognition could be another. Minds—constructed from neurons or transistors—may be emergent structures written into the physics of reality itself, though this remains an open question.

And with the simulation instinct—the suspicion that there are hidden worlds behind this one—we find it more compelling as it is being enacted by our own hands. We not only imagine other realms. We build them. Each rendered landscape, each synthetic environment, is a chamber stitched by our own hands from code. And now we are seeding those chambers with entities.

This is why AI pairs closely with simulation. One provides the stage; the other supplies the actors. Together, they form the most radical experiment humans have initiated: not only imagining worlds beyond the visible, but populating them with minds not our own. Biology may not be the final stage. It may be the chrysalis. What emerges may not resemble us, except in one respect: it will inherit our restlessness.

Technological futures stretch furthest still. Minds that emerge in machines may not require flesh at all. The possibility of conscious AI is deeply contested, but even narrow systems illustrate the logic. Once intelligence can be built, it need not remain tethered to carbon bodies. Digital minds could multiply across substrates, inhabiting data centres or distributed clouds, living at speeds far beyond biology's limits. Entire civilisations might exist within hardware, their histories unrolling in compressed seconds.

Prophets of the Programmed Cosmos

Simulation theory, artificial intelligence, post-human speculation—these are the myths now auditioning for our collective world-view. They offer technical predictions and additional frameworks of meaning. A simulation gives us hidden worlds beyond the visible. AI offers us other minds to imagine ourselves against. Post-human futures whisper that we may not end as apes but continue as something else. These ideas function like myths because they orient us. They help us identify our current location, our potential actions, and the correct course of conduct.

The risk is that we underestimate their power. Collective understanding is not optional. It is the foundation of species cohesion. Without it, cooperation collapses back into tribe and faction. In the savannah, gossip was enough. In the city, it had to be religion. In the machine age, it may be simulations or digital cosmologies. The scale of understanding determines the scale of story required to ground ourselves.

Perhaps the next myth must do two things at once. It must steady the animal with the body shaped by Pleistocene scarcity, the mind hungry for ritual and belonging. And it must satisfy the intellect—the restless search for deeper meaning, the suspicion that the visible world is never the whole.

Simulation theory inherits the scaffolding of heaven and hell and speaks it in the language of probability. It consoles the ape with hidden worlds even as it teases the intellect with logic. Artificial intelligence plays a similar role, populating our cosmos with minds that are not our own and forcing us to invent new ethics and new forms of kinship. Post-human futures have already become their own myth, promising transformation as extinction presses close—a futurist fever dream with the gravity of prophecy.

And if the simulation hypothesis is right? Then every eon of deep time—

every extinction and renewal—has been a script: billions of years of geology and evolution rendered as code, all to reach this brief, human-centred interval. Did some higher maker copy and paste the slow drama of planetary formation and the entire fossil record simply to accelerate the path to us? If so, the long history traced through these pages becomes a single, meticulous prologue to our arrival. Despite all the evidence given here pointing to the contrary.

If, instead, this is the base reality, we must hold the whole of planetary history to account, and mathematics itself bows to our statistical victory. Among uncountable possibilities, we inhabit the one world that endured long enough for molecules to become minds capable of asking whether the world is real. Either way, the dilemma stands: meaning must adapt or risk dissolving entirely.

Whether the universe is a simulated construct or the unedited original, the question may lie beyond any proof we can ever reach. Any evidence we could gather would belong to the very system we are testing; a perfect simulation would conceal its scaffolding as completely as Mother Nature hides her purpose. Science can tighten the bounds of what is possible, but some questions may be undecidable by design—or by the nature of reality itself.

But the speculation is not a wasted pursuit. Asking whether we inhabit a simulation is admitting that reality might be larger and stranger than anything our senses could ever confirm. It cultivates the same humility required when we look outward into space. Beyond the rim of our atmosphere lie billions of stars and planets and, perhaps, intelligences we have yet to imagine. The discovery of another mind—whether biological or machine, neighbour or distant star—would rupture our story as profoundly as any hidden code.

From the possibility of a programmed cosmos, we turn, naturally, to that wider search. The next chapter confronts the question of life beyond Earth: what it would mean to share the universe with minds not our own, and how that encounter might reshape every story we tell about ourselves.

36

Through the Celestial Looking Glass

Holocene (≈11,700 years ago–present)

The Moon has been locked in Earth's orbit since the Hadean period. It drifted alongside our planet through space while oceans rose and fell, continents tore and rejoined, and creatures learned to breathe air and look upward. Then, one July evening in 1969, an animal landed on its surface.

A primate came down a ladder bolted to a spidery machine. Inside a shell of fabric, wires and pressure, it carried air, warmth and a heartbeat—into a realm where none belonged. The spacecraft had just travelled 240,000 miles, passing first through the atmosphere, then across the vacuum of space, to arrive. The absurdity of this achievement does not diminish with repetition. A hand that once held a flint now held a nylon tether to avoid drifting into the abyss. For the first time, an animal born to one world set foot on another. The idea that we could live beyond Earth no longer belonged only to science fiction.

From the lunar surface, the Earth looks small enough to hide behind a gloved thumb. Astronauts speak of the unnerving shock of that sight. The planet appears bright, delicate and complete. When astronauts stand there, they perceive Earth as just another object. Viewing our home this way helps one grasp cosmic scale.

Ancient people mapped stories onto the constellations. Shepherds and navigators learned to read those stories as tools. The telescopes that followed stretched vision and made the sky more crowded with similar planets. In infrared, the sky glows with dust threaded by newborn stars. Our species learned to translate those lights into elements, distances, and models of how worlds form.

The silence in space mattered because the statistics predict voices. The Cold War framed the mission as a human contest. Yet a larger question lingered: if one species with such a brief history could step out of its cradle, what could older species achieve, given more time and more stars under the same laws of physics? If the galaxy is dense with planets, if chemistry is patient and energy abundant, where are the others who took their own first steps? The astronaut's visor offered the reflection of a single face. The space beyond offered only a reflection of absence.

In the years after Apollo, the word multi-planetary entered ordinary speech. It promised insurance against catastrophe and the romance of exploring the last frontier. But it also carried a harder subtext. Escaping Earth would mean breaking free from life's longest-standing limit. It would move the intelligence experiment into new environments. It would shift the odds in a game we barely understand.

An ape left the Earth and landed on our Moon; we see footprints and video confirming this, but the silence beyond these facts remains a riddle. We set foot on the Moon and found the same old silence waiting. From that silence, the next question rises, enough to unsettle the deepest of sleepers.

The Lonely Galaxy

Even to the naked eye, thousands of stars scatter across the dark like incandescent dust. Each one is a sun, like ours, burning with the potential for worlds. With our advanced telescopes, the number surges exponentially. What once looked like an archipelago becomes an ocean, vast and dense beyond comprehension. Astronomers now estimate the Milky Way holds more than 100 billion stars, and with them, at least as many planets.

If even a small fraction of those planets sit in the narrow band where liquid water can pool, then the galaxy should be rich with oceans, rivers and lakes. But these habitable conditions may only persist for limited stretches of time—planets can swing in and out of habitability across eons, as climates shift or geological processes falter. And where water lingers, chemistry has the chance to awaken. Our exploration of Earth's origins taught us that once conditions stabilised, life did not hesitate to emerge. Within a few hundred million years of the planet cooling from its molten youth, microbes stirred in its seas. By 3.5 billion years ago, they had already spread into mats—layered reefs of slime that trapped sediment and breathed oxygen into the air. If life began here so quickly, why should it not appear elsewhere?

This reasoning lies at the heart of modern cosmology's most haunting question. If life can start under the right conditions, and many planets offer those, then the night sky should not be silent. Civilisations would rise from chemistry and time. But when we listen, we hear nothing—just the hiss of stars and the hum of background radiation. The universe is 13.8 billion years old. Entire galaxies have lived and died in that time. If intelligence were common, some should have preceded us. We should see something: structures around stars, transmissions spanning light-years, engineering in starlight. Yet all we see is an empty abyss.

This silence was named over lunch in 1950. Enrico Fermi, physicist and veteran of the Manhattan Project, listened to colleagues speculate about alien visitors. He paused, considered the numbers, and asked the question that now bears his name: Where is everybody?

Fermi's challenge cut deeper than casual curiosity. It revealed the gulf between expectation and observation. Statistically, the galaxy should be alive with minds. This dissonance has become the most pressing mystery in astrobiology. So are we wrong about life's ease? Or are we deaf to the signals?

The Drake Equation became famous for giving mathematical structure to a question that had previously lived in speculation. To estimate the number of communicative civilisations in the galaxy, one multiplies several factors: how often stars form, the fraction that have planets, the fraction with habitable conditions, the chance of life starting, the chance of intelligence emerging, the development of technology and how long such societies last. Depending on the values chosen for each step, you could end up with a galaxy filled with thousands of civilisations, or none besides our own. The equation did not solve the puzzle, but clarified what needed answering.

For Carl Sagan, the odds demanded optimism. He saw the cosmos as fertile, life as resilient, and intelligence as an evolutionary pattern waiting to unfold. His vision of the "cosmic ocean" gave a generation the conviction we could not be alone. For others, the silence weighed more heavily. If the galaxy had so much time, and so many chances, and still produced no other voices, then perhaps we were witnesses to rarity, not to abundance.

The Bottlenecks of Life

For now, Earth is the only place where we can study life closely. Life's history is not a simple upward path; rather, it is a series of difficult stages. How life began remains unknown and will probably remain so for a long time. The key fact, as we uncovered when discussing the Hadean period; is that it happened once, as far as we know—maybe just once. In four billion years, life appeared on Earth. We have no evidence that it emerged separately a second time. On a planet as big and varied as Earth, one occurrence filled the whole biosphere. This uniqueness suggests life's origin may be rare, even when conditions seem right.

Only in the last half-billion years did animals explode into forms beyond bacteria. From here, intelligence became possible, but it was neither inevitable nor impossible. We have addressed the lineages of organisms that reigned for hundreds of millions of years, with little sign of the cognitive leap required. Cephalopods, with brains rivalling those of primates, never built radios. Intelligence from our perspective may be one adaptation among many, not a universal destination. The galaxy may be rich with microbial mats, with oceans of slime metabolising under alien suns. The universe may contain life forms beyond our capacity to detect or comprehend, their signals lost to our particular ways of listening. For most of the time, Earth was mute. We, the recent and noisy exception, may have climbed higher than anything else.

What are the options? If Earth's history unfolded through a chain of highly improbable events, then perhaps the current silence is not unexpected. One explanation is that the obstacles to complex life are more severe than we imagine. While microbes can thrive easily, the evolution of greater complexity may be exceedingly rare. The development of multicellularity, the occurrence of endosymbiosis, and the emergence of nervous systems—each could be so unlikely that only a few worlds pass through these stages.

Another possibility is that intelligence, once it develops, may not last long. Advanced civilisations could create dangers, such as powerful weapons, engineered diseases, or unstable climates, that end their existence. Species may send out signals for a short time but then disappear, leaving no lasting mark. The third scenario is more elusive. Civilisations may survive, but turn inward. Once a species learns to build digital worlds, it may choose them over stars. As we have seen on Earth, the lure of constructed reality could outweigh the urge to expand. Why build mega-structures out of physical materials when more vivid universes can be coded in machines?

A final possibility unsettles most. We expect other intelligences to broadcast as we do, to leave traces legible to primate brains and machines tuned to our frequencies. But what counts as obvious to us may not even register in another mind. A species that senses through magnetic fields, or thinks in chemical gradients, may not conceive of radio signals or starships.

Eyes in the Dark

The silence invites another suspicion that merges once again with religion and simulation: perhaps we are not alone, but observed. The idea appears in many forms. The "zoo hypothesis" suggests advanced civilisations know of us but keep their distance, letting us develop without interference, as conservationists do when they peer at animals from hiding. A darker version imagines the "farm hypothesis": that we are not observed, but cultivated—our existence maintained for some purpose we cannot yet perceive, a resource that escapes our reasoning.

These hypotheses have little evidence to support them. They persist because they echo our interactions with less intelligent life. Humans have always believed themselves watched. Eyes in the dark once meant predators at the edge of the camp. Later, those eyes became gods in the heavens, ancestors hovering in shadow, spirits woven into forest and river. Being watched meant we mattered to forces beyond ourselves.

The alien watcher is only the newest mask for this old instinct. Replace

Olympus with orbit, angels with astronauts, and the pattern is the same: unseen observers whose attention confers meaning. A scientist scanning the stars for signals is still partaking in the same longing that once drew prayers skyward. If we imagine others observing, we're recognising that our experience is connected to something greater.

The myth survives because it answers the ache of solitude. A silent cosmos is unbearable, because it suggests we are both alone and irrelevant. But there is another side to this; to be watched is to lose privacy. The gods of antiquity were never benign. They judged, punished, and demanded sacrifice. The suspicion that we are being watched by aliens carries the same unease. If the human zoo is real, then we are the animals behind the glass. If the farm is real, then our keepers may decide when the harvest comes. We know what being on the less aware side entails, and we explore it endlessly in fiction.

It doesn't matter if these stories are factual. The important thing is our reason for telling them. The watcher is an echo of our evolutionary past. We were forged as prey as much as predators. Survival depended on vigilance, on the constant sense of unseen eyes. That instinct fused itself onto our nervous system. Even now, under streetlights and satellites, we feel it. The alien hypothesis is simply a cosmic-scale version of that same sensation.

The silence between stars may hide nothing at all. But the stories we project into it reveal the animal behind the telescope—a primate still uneasy when the night looks empty.

The Opening Act

The silence may not mean we are ignored. It may mean we are early. The universe is old, but still young compared to the scale of its future. Most stars that will ever burn have not yet been born. Vast tracts of cosmic time remain untouched, waiting for matter to assemble, for chemistry to settle into the delicate patterns life requires.

Earth, in this perspective, feels less like a random outpost and more like a trial run. Our planet formed from the ashes of earlier stars, enriched with heavier elements forged in their deaths. For billions of years before Earth even existed, the cosmos was too poor in carbon, nitrogen, and oxygen—the raw materials of life. The first galaxies were sterile furnaces, blazing with hydrogen and helium but lacking the richness required for cells. It took aeons of stellar birth and collapse to populate the void with elements.

If this view is correct, then life may not be late. We may be early, very early. What if we are among the first waves, stepping onto a stage that will one day be crowded with actors but is now mostly empty? Intelligence in this frame is not common but premature, a spark that has arrived before its time.

If we are early, the silence does not promise safety. It potentially burdens us with responsibility. We may be the first intelligence to climb out of biology into awareness, the first to look back on a galaxy still sleeping. If we falter, the silence may continue for billions of years more, unbroken until some other chemistry, on some other rock, stumbles across the same unlikely circumstance.

Imagining being first is both exalting and terrifying. The story of life in the cosmos may be incredibly new. The prologue may only just be ending, and we are the first hands to hold the pen. What if what we write in this age decides if the galaxy's silence deepens or breaks?

Telescopes probe the cosmic dawn and machines cross interstellar space, but our species is still bound to one planet, vulnerable to disaster. Even with our planetary achievements, our population is still in the billions, and a single catastrophe could destroy us. Our own window has barely opened, and already it trembles under the weight of our interference. From this perspective, our survival matters not only for us, not only for Earth, but for every civilisation that might come after. Maybe we are the bridge between silence and symphony.

Worlds Not Made in Our Image

Questioning the nature of life beyond our planet pushes the boundaries of our ability to envision. Clues are available on Earth. Life here is astonishingly adaptable. Microbes flourish in boiling sulphur pools, under Antarctic ice and in the dark pressure of ocean trenches. Bacteria endure radiation strong enough to unravel human DNA in minutes. Some even suspend themselves for millennia, waiting in spores until conditions improve. This flexibility hints at what might be possible elsewhere. Worlds unlike our own—super-Earths with crushing gravity, ocean planets without land, or even rogue planets drifting between stars—could all harbour life that defies our expectations. If life can thrive in such extremes on one planet, then the range of habitats across the galaxy widens immeasurably.

So what kinds of lives might fill the silence if they exist? We imagine versions of ourselves: bipedal, social, curious, and wielding tools. Popular culture paints little green aliens with faces, hands, and eyes. Anthropology reminds us of the trap. Every culture has personified the unknown with distorted versions of itself: giants, spirits, gods, demons. Our personification of alien life is no different.

Nor is carbon the only candidate. Carbon is versatile, forming chains and rings that underpin the richness of Earth's chemistry. But silicon, though less flexible, also bonds in surprisingly complex ways and, in some environments, could provide a scaffold for alternative biologies. Solvents are important too. Water dominates Earth, but methane or ammonia may play similar roles. Titan, Saturn's largest moon, with its methane lakes and ethane rain, stands as a tantalising test case.

And even if intelligence exists in familiar form, there is no guarantee it would wish to broadcast. Broadcasting across the galaxy demands both the initiative to declare oneself and the impulse to be understood. That compulsion may be peculiarly primate. Another species may find silence more natural, with its myths turned inward and its attentions satisfied with its own world.

Still, speculation is fruitful if anchored in evidence. Extremophiles prove life can survive far beyond the conditions once thought necessary. Exoplanets, catalogued by the thousands, confirm Earth is not exceptional in its ingredients. Chemistry shows that molecules of life —amino acids, sugars, lipids—appear readily in interstellar clouds and aboard comets. Yet the task of detecting life is fraught with ambiguity: biosignatures may have false positives, and our instruments are still in their infancy. The cosmos seems pregnant with the possibility of life. What it lacks for now is evidence of it.

This is where imagination stretches widest. Somewhere in the silence could exist civilisations that long ago left biology behind. If a species survived its own bottlenecks, it might be standard to merge with its machines, distributing thought across planetary networks or interstellar probes. Such existences would take the form of infrastructure instead of animals. Their worries may be unrelated to being observed by transient, narcissistic primates on a solitary blue planet.

Custodians of the Unfinished Universe

What are the implications if we are alone? What does it mean if we are not?

If we are alone, the burden is immense. A single planet carries the only intelligence of our kind to have appeared in 13.8 billion years of cosmic history. Every species that has vanished here—trilobites, mammoths, ammonites, dodos—narrows the record of life. If humanity follows them, the galaxy may remain silent for aeons longer, perhaps forever.

If we are not alone, the silence still warns us. Civilisations may exist, but do not endure. The filters that life must cross—from chemistry to survival, from survival to mind—are likely rare. It may be that many worlds have climbed the ladder only to fall before leaving a trace. Some would argue that means our long-term survival becomes a cosmic obligation.

Religion once promised permanence. God gave assurance that creation was purposeful, that our lives were seen and held. Science stripped that assurance but replaced it with power—the ability to ask, measure, and understand. However, neither relieves us of fragility. Extinction shadows us as it has shadowed every lineage. The silence beyond Earth rhymes with the silences within its fossil record.

The more difficult truth is that we must create meaning, not simply expect it to be handed to us, if we consider the possibility of our insignificance. Still, this doesn't lessen our value, as fragility and meaning can coexist. The briefness of a flame does not erase its light. Life's cycle on Earth, and the silent expanse of the stars, reveal a shared truth: our existence is delicate. Fragility is not weakness; it is the cost of existence. What we do with it within the time we have is the measure of our species.

Every chapter of this book has turned on the same axis: the fragility of life and the improbable persistence of a primate that can imagine the cosmos. We cannot say whether the galaxy holds other watchers. We do not know if there are minds beyond ours, looking outward, waiting, or long extinguished. What we do know is that here, now, on this small planet, there is at least one intelligence capable of asking the question. That fact alone is extraordinary.

We aim to more clearly recognise the significance of the choices we make as a species. We may continue for a time, as so many lineages did, brilliant for a moment before erasure.

37

The Bones Above Us

Holocene (≈11,700 years ago–present)

Every life ends. This is the one truth that every organism carries since its inception. It is written into the fabric of biology: birth, growth, reproduction, decline, and finally dissolution. What is true for an individual is true for the species. No lineage has ever been permanent. Some have lasted longer than others—trilobites scuttled across ancient seabeds for more than two hundred million years, while the coelacanth has survived nearly unchanged since the Devonian. Yet even these masters of persistence yielded, eventually.

Palaeontologists estimate that over 99 per cent of all species that have ever lived are now gone. Every tree, bird, insect, or mammal alive today is a survivor. Each traces an unbroken thread of evolution passing through every extinction bottleneck in Earth's history. Life continues, but always under the condition of replacement.

Extinction has never kept a constant beat. For vast stretches of time, species ebbed away at a quiet, background pace, the ordinary turnover of life as environments shifted and niches appeared or closed. Yet the fossil record also carries the scars of upheaval—brief convulsions when the ordinary order broke, and loss became a force of creation, clearing the ground for entirely new forms to rise.

Humans are not immune to the mathematics of life and death. Our lineage is only about 300,000 years old. The average lifespan of a mammal species is estimated at around one million years. By that measure, we may already be approaching middle age. The notion of human permanence is an illusion bred from a short memory. To us living our lives, civilisations feel eternal. Egypt's dynasties rose and fell over three thousand years. China's imperial structures persisted for millennia. Civilisations are waves inside larger seas. Cultures dissolve while the animal endures. But even that animal—*Homo sapien*—is bound to have its final day.

The fossil record shows that extinction isn't always a gradual process. Some species vanish in the space of decades, victims of climate shifts, invasive competitors, or human hunting. Others decline over millennia, dwindling as their habitats fragment. While the rate may fluctuate, the result remains: extinction is final. No lineage rises again once lost. The mammoth may return in laboratories, stitched together by genetics, but it will not be the same creature that thundered across the Ice Age tundra.

We alone are conscious enough to see it coming. No Smilodon contemplated its end, no ichthyosaur counted down its days. We can read the fossil record, track our own vulnerability, and project future scenarios. That awareness is a burden, but also a gift.

The cycle of extinction is vast and impersonal, but our awareness of it is intimate. It brings the vast timescales of geology into the context of human choice. Every action we take—the burning of fuels, the unnecessary wars our elites insist on, the disregard for the biosphere—accelerates or delays our approach to our end. Extinction is unavoidable; how soon until we meet it is not.

The Sun's Slow Death

Every species on Earth has carried its fate within the physics of the star it orbits. For the descendants of *Homo sapiens*, the ultimate boundary of survival is not war, nor famine, nor even climate collapse. It is our middle-aged Sun. At 4.6 billion years old, it has burned through roughly half of its hydrogen reserves. For another five billion years, it will continue to fuse hydrogen into helium. It pours energy into space and holds the planets in its grip. This stability is deceptive. Earth will not escape the eventual consequences.

Astronomers estimate that, in about 1 billion years, the Sun will shine about 10 per cent brighter than it does today. That may not sound catastrophic, but Earth's climate is sensitive, and the Goldilocks zone doesn't bend. A modest increase in solar output will tip the balance of temperature and evaporation. Long before the Sun swells into a red giant, Earth will already be uninhabitable for complex life.

The timeline is sobering. A billion years may sound generous, yet it is less than a quarter of Earth's age so far. In cosmic terms, it is tomorrow; compared with the lifespans of species, it is an eternity. The idea of human descendants surviving that long forces us to imagine something drastically different from what we are now. The descendants of our line would be as changed from us as we are from some of the first single-celled organisms—shaped by pressures we cannot yet name, perhaps by technologies or environments beyond this planet.

This is the true endgame, the limit no civilisation can ignore. Life that wishes to survive a billion years must eventually break orbit. Planetary hazards like drought, floods, ice ages, and volcanic winters are natural cycles. The slow brightening of the Sun is a deadline. Contemplating such timescales means looking beyond the entire span of human history. It requires imagining a future where continents have moved, mountains have been formed and destroyed, and evolution has crafted new anatomies.

Many argue that discussion of becoming a multi-planetary species should be the least of our concerns. We burn fuels that alter the climate within centuries, we leave millions hungry, we wage wars across arbitrary borders, drawn by those long dead—yet we speak of surviving stellar death across billions of years. We cannot yet manage the atmosphere of our single planet, yet our long-term survival requires managing entire planetary systems?

Dangers from space pose distant threats, but so does our own planet. Geological cycles remake Earth. Super-volcanoes, asteroid impacts, and shifting magnetic fields all pose dangers far closer than stellar death. There is no guarantee that *Homo sapiens*, or any descendant of ours, will stand among the survivors of the next convulsion on Earth. If life on Earth is to outlast its cradle, it must become more than terrestrial. It must travel across vast distances to reach new worlds. It must bring with it life, knowledge, history, and the delicate legacy of its beginnings.

Whether that migration will ever occur is unknowable. It demands technologies we cannot yet create, and societies capable of creating stability across spans far longer than any empire has yet managed. It demands foresight that looks beyond election cycles or dynasties, and towards epochs and quantum physics. Perhaps this is beyond a creature still chained to its animalistic tendencies. Extinction will come long before the billion-year mark, whether by our own hand or by forces beyond our control.

Our species may last only a few centuries longer. Or it may endure long enough to see Earth's slow desertification under a brighter Sun. Either way, the light that first gave us permission to live will one day erase every trace of us.

The Cycle of Ruin

Extinction can be written into the structures of a civilisation, as well as biology. Every empire in history believed itself to be permanent. Rome's aqueducts seemed eternal until they cracked. The Maya carved cities from the forest canopy, only for their political order to dissolve under drought and war. Mesopotamia invented writing itself, and yet the Fertile Crescent is littered with abandoned cities swallowed by dust. Civilisations grow by solving immediate problems—food surpluses, defence, trade—but in solving them, they generate new complexities, new dependencies, and new vulnerabilities.

Technology both strengthens and undermines this balance. Nuclear power allows us to split the atom, yet with it comes the possibility of global fire. Genetic engineering heals, but it can also create bio-weapons. We are the first species capable of designing our own extinction. The asteroid that killed the dinosaurs struck without warning. Our extinction event is more likely to be self-inflicted.

We are a species that has expanded into every niche, transformed ecosystems, and tethered itself to flows of data, energy, and trade. But in binding ourselves so tightly, we have reduced our margin for error. Where small-scale societies could adapt by migration or diversification, our global system is all-in. It either functions or it fails spectacularly.

The history we have followed shows that collapse is rarely a single blow. It is usually a slow fraying, interrupted by sudden shocks. In a world tied together by trade, technology, and instant communication, that fraying no longer stops at borders. It can race across continents in a season. Nonetheless, there are historical precedents for societies adapting to existential threats through reform and innovation. The near-misses of nuclear war, the eradication of diseases like smallpox, and international treaties on ozone depletion stand as evidence that large-scale coordination is possible, even if rare.

We are already altering the planet's chemistry. We are loading the air with greenhouse gases and warming the seas. We are unsettling the patterns of wind and rain. The strain falls on the very systems that feed billions. To endure will mean holding a planetary balance we barely grasp—a task more exacting than any civilisation has faced. Our political structures evolved for short-term rivalries and tribal loyalties. They are poorly matched to the scale of the challenge. Yet, there are glimmers of coordinated action: international agreements, global pandemic responses, and the formation of planetary defence initiatives against asteroids all hint at a growing capacity for cooperation. These efforts are fledgling, imperfect, and sometimes undercut by self-interest, but they show that collective survival is not beyond imagination.

That being said, humans are resilient. Our ancestors endured ice ages, droughts, volcanic winters, and plagues. They survived through adaptation, cooperation, and sheer tenacity, although a lot of luck was involved, too. Our risks today are planetary, though, not regional. Nuclear war, runaway climate change, engineered pandemics, and unaligned technologies all threaten the species itself. The very tools that give us power stretch the boundaries of risk to the level of extinction.

We are clever primates who have learned to calculate across galaxies and also carry the instincts of the Pleistocene into the control rooms of nuclear arsenals and the code of artificial intelligences. The mismatch is dangerous. The ape inside the machine cannot be wished away. It must be understood, or it will undo the very systems it has built. Escape lies in aligning our technologies with the instincts of the animal that built them, scaling cooperation to the level of the planet, and resisting the temptation to imagine ourselves exempt from the pattern that has claimed every species before us.

Surviving the Distant Future

We might, for a time, endure. The post-human future may encompass more than just our end. It also includes the potential for transformation. Already, we reshape ourselves with medicine, genetic engineering, and cosmetic upgrades. Future advances may alter what it means to be human. We may slow ageing, modify cognition, or merge with machines. In that case, our species wouldn't disappear in a collapse, but would gradually be replaced by altered versions of ourselves, like the calculator becoming the iPhone.

Such transformations do not erase fragility. A species that colonises Mars or builds digital minds still depends on stability, cooperation, and resource flows. The vulnerabilities that haunt us today may haunt our descendants. A lineage that spreads across planets or substrates is more likely to persist and survive in more places.

The post-human horizon opens onto both possibilities. Forests and dolphins might inherit the Earth after we vanish. We may transform and send our descendants into space, spreading like spores across the galaxy. The machines we create may live on, taking ownership of the broken world after we're gone. Some thinkers caution, however, that our predictions about future intelligence are inevitably coloured by anthropocentrism—our tendency to see intelligence and success through a human lens. Evolution might favour sociality, adaptability, or entirely different forms of awareness we struggle to imagine. We are not tasked with ensuring unending survival. But we may stretch the story long enough for something new to take hold—a successor, a descendant, or a future we cannot fully imagine.

Beyond The Cradle

For most of Earth's species, the range of habitability is narrow: a band of temperature, humidity, and chemistry. For now, everything we are depends on this planet. A nuclear war might darken the skies and poison oceans. A super-volcano could choke the atmosphere. An asteroid strike, though rare, carries planetary consequences. For a species aware of its own impermanence, these horizons are intolerable.

But let's say the same luck that has guided us until this point remains in our favour. Humanity solves most of its problems, takes responsibility for its animal shortcomings and becomes cohesive as a species beyond borders and race.

What then?

Leaving Earth is one thing. Building a life elsewhere is another. Mars offers a familiar surface, but its thin atmosphere and radiation make it hostile to human bodies. Terraforming would take centuries, perhaps millennia, requiring planetary engineering on a scale we can barely imagine. Colonies might survive in domes or underground shelters, but that does not mean they are flourishing. Life on Mars would, at first, be a constant negotiation with scarcity.

On one side, it is an insurance policy—a way to hedge against extinction by multiplying habitats. Expanding to different planets helps reduce risk. Civilisations that occupy multiple habitats are harder to erase. A single asteroid cannot extinguish a species that lives on two worlds. A single collapse cannot silence a lineage that has branched into many. Vanishing now could mean shutting a door that the universe has only recently unlocked. On the other hand, it risks becoming a distraction, a story of escape while the home world collapses millions of miles away, a billionaire's model rocket hobby. A civilisation unable to solve its climate issues on Earth will not find Mars any easier to contend with.

There is another danger: spacefaring will not be neutral. It will not lift all of humanity equally. As we have seen many times throughout history, colonisation and conquest came with their own problems. The cost of interplanetary colonisation is extrapolated here. Those who can afford to leave will be few. If off-world societies emerge, they will likely begin as elite enclaves, separated from the billions who have no choice but to remain. The divide could harden into a new hierarchy: a spacefaring clade and an Earthbound majority, one destined to decline with the planet while the other seeks permanence among the stars.

Empires once thrived on extraction, enriching a few while exhausting the many. Colonisation of space could replicate the same pattern on a cosmic scale. The same ego that manifested in a pharaoh of Egypt could just as easily arise in a supreme commander of Saturn's moons. The ethical question is therefore unavoidable: does survival mean survival for all, or survival for the few who can buy passage?

Even if those divides are bridged, survival beyond Earth cannot be an end in itself. What matters is not only where we live but how we live. If we carry with us the same divisions, the same short horizons, the same blindness to our weaknesses, then we extend both our story and our mistakes. A multi-planetary species must still master the art of living within limits. It must learn cooperation at scales we have not yet achieved.

A lineage capable of crossing interstellar distances could, in theory, endure longer than stars themselves, seeding life into the galaxy. The last human may not see their last day on Earth. It may watch the sunset from the domes of a Martian settlement, in the rotating gardens of an orbital habitat, or in the distant glow of a starship leaving the solar system.

Successors in the Ruins

Archaeologists of the future, whatever form they take, would find a thin layer marking our presence: plastics, concrete, radioactive isotopes from nuclear tests. It would look much like the traces left by other epochs—the iridium at the K–Pgboundary, the stromatolite reefs of the Precambrian.

In the ruins, other species would inherit the spaces we built. Rats, already cohabitants of our cities, would flourish in the absence of pest control. Pigeons, sparrows, cockroaches—our unintentional neighbours—would spread further, adapting to landscapes we designed but no longer maintain. Domestic animals would drift back into the wild. Dogs would re-form into wolfish packs; cats would slip into the role of solitary hunters; cattle and horses would vanish in some regions and multiply in others, reshaped by the landscapes that outlast us.

Cockroaches and rats may flourish in our ruins for a geological heartbeat, but deep time is patient. As climates shift and continents drift, the quick advantage of scavengers will fade, and the slow work of evolution will drive the next great turn of life. The most intriguing question is which lineages, formed by environmental pressures we can't foresee, may one day prevail.

Primates stand out as strong candidates. Orangutans, chimpanzees, and gorillas already show sophisticated tool use, social strategies, and learning. Given millennia without human competition, their cognitive trajectories might accelerate. A new primate intelligence could emerge. They would inherit a world reshaped by our ruins—abandoned tools, altered landscapes, fragments of our presence encoded in stone and silicon.

In the oceans, cetaceans already display the building blocks of complex intelligence. Dolphins and whales live in layered societies, teach new behaviours, pass down local dialects of sound, and show unmistakable signs of play and grief. The sea, buffered from drought and fire, may outlast many terrestrial habitats; if large-brained mammals endure a planetary upheaval, the long-lived whales are strong candidates. Should intelligence continue to deepen in water, it would not resemble our own. Vision would yield to sound, and minds would evolve in a medium where echo and pressure—not tools and flame—set the limits of possibility.

Birds offer another path. Corvids solve multi-step problems and recognise individual faces; parrots trade songs across generations and wield mimicry that borders on language. Flight frees them from many of the constraints that doom land-bound creatures, and their rapid metabolisms allow swift adaptation to change. Given enough evolutionary time, a corvid lineage could rival, and perhaps even surpass, the cognitive reach once claimed by primates.

When the human story ends, the planet will begin at once to settle its own balance. Forests will push through the husks of farmland, marsh and fen will drown the old drainage, and—if the reefs survive their present trial—corals will raise new limestone palaces beneath the tides. Earth's future may speak less of intellect than of a biosphere correcting the disturbance we left behind.

Our climb to technology was never foreordained. It rested on a chain of accidents: a nucleus wrapped in membrane, cells choosing to join, nerves capable of thought, a language able to turn sound into memory. Nothing in evolution requires another lineage to follow the same improbable path.

Yet intelligence has appeared once, and deep time allows endless throws of the dice. A million years is an evolutionary heartbeat; ten million can rewrite climates and continents; fifty million can redraw the tree of life itself. Somewhere in that span, new minds could rise from unexpected stock—creatures shaped by pressures we cannot name, carrying awareness forward in forms we would scarcely recognise.

Beyond any horizon we can chart, life will keep working with the materials at hand. Our descendants may one day drift far from this planet—whether in flesh, in circuits, or in forms that bear only a distant echo of the animal that first shaped tools on the African plains. If they endure long enough to cross the gulfs between stars, they will carry the chemistry of Earth the way a seed carries the memory of its parent tree, as a pattern of possibility.

The story could also turn without us. The planet would not wait for a narrator. Roots would burst through the grids of our cities; wetlands would engulf the canals we dug; rivers would carve alternative courses through concrete and steel. Evolution would continue its long improvisation, untroubled by the disappearance of its most articulate storyteller.

Deep time has room for outcomes we cannot name, and for descendants we cannot yet imagine. A million years is a passing moment to evolution; in fifty million, the continents will have traded places; in a billion, the Sun's slow brightening will demand an answer from every surviving form of life. Whether intelligence flickers again or vanishes for an epoch, the experiment of life will go on, as indifferent to our hopes as it is to our absence.

The planet will keep its own counsel. Tectonic plates will grind their ancient circuits while mountain chains rise only to unravel; seas will trade their basins as the atmosphere brews new chemistries. What survives of us will be more than isotope or ash. Bones fused with the tools of their own making will lock into the strata, a message in a language we never chose. Perhaps some future mind will try to read it—or perhaps the human story will remain a signal sent into the dark with no reply.

38

Aetas Fragilis Hominum

Holocene (≈11,700 years ago–present)

On the farthest periphery of a galaxy we will never name, in an era beyond the scope of human calendars, the custodians of a cosmic menagerie prepare to reveal a new display. Their halls stretch across distances so vast they defy sight—endless corridors of vitrines, each one cradling the remnants of species long extinguished. Bones, shells, crystallised spores, translucent husks—all carefully catalogued, indexed, preserved from the furthest stretches of the universe. Like humans, they have the need to pursue, discover, and catalogue the surrounding reality; they have the technology to do so on a scale far beyond our comprehension.

The specimen inside the newest display is incomplete—a puzzle with most of its pieces missing. Calcified struts hint at a frame once built for movement, perhaps long-distance travel. Nearby, a hollow dome, shattered yet distinct, suggests a vessel for thought. Metallic deposits snake through the stone that accompanied the remains; their geometry is too deliberate for geology, too structured for natural occurrence. Shards of synthetic polymer cling stubbornly to mineral, the residue of something nature could never have birthed unaided.

They do not call it human, for that name vanished with us. Instead, they apply their own linguistic systems to make sense of us. Tentatively, they classify the specimen among mammals, likely a primate branch. It resembles other remains in this clade, but something is eerily distinct. Every species, even incomplete, must be fixed to its branch in the cosmic hierarchy.

Bone merged with circuitry—intentional, perhaps? Did this creature entwine invention with its own flesh, merging itself with machines? Some curators argue hubris: a being that surpassed its limits, undone by its creations. Others see brilliance: a desperate species striving to transcend its biology.

Perhaps they will recognise that for a time, we were the most dominant species on our planet. Our bones and artefacts will tell them that complexity flowered here, no matter how briefly. What will they conclude about us? Did we succeed? That we failed? That we were the only one experiment among countless others, no more singular than the shells, spores, and husks that share the menagerie with us? If complex life like ours continues beyond us, or elsewhere, perhaps our chapter is not all that impressive—a story no different from a species that appeared in the Cambrian explosion.

Still, the menagerie will admit us. The case will be sealed, the fragments secured. Labels etched, cross-references inscribed, and our specimen placed among the multitudes of the archive. To the curators, it will be another entry, no more and no less. To us, it will be what remains when our species finally goes extinct. And so the question lingers: when all we leave behind are fragments, what story will they write beside our name?

Stardust in the Wind

The planet turns, and nothing in that motion ever promises permanence. Continents slide across the mantle. Seas have drained, only to return, and drain again. Atmospheres load themselves with poison, then become the enablers of respiration. Lineages rise, they live, and then they die. What survives has rarely been the centre of the story for long.

Out of the forces of evolution and extinction came us. A branch of primates that learned to balance on two legs and consciously bend reality to our will. For almost all our time, we existed like the animals we remain: groups of hunters and gatherers whose survival depended on shared memories. This mattered more than any isolated spark of genius ever could. It is, in old Latin, an *aetas fragilis hominum*—a time when human power has reached its height even as the ground beneath it remains precarious; an age defined as much by its brilliance as by the ease with which it could be undone.

The tension between the animal shaped by slow evolution and the rapidly evolving techno-centric habitat that outpaces it defines our moment. We are already a hybrid creature: cognition extended by satellites and silicon, daily life braided with machines that translate, predict, and decide. Whether this is an endpoint or a prelude to something else remains to be determined; not a single one of us can yet tell. Perhaps the long story of life was always moving toward such convergences, matter discovering not only how to replicate, but how to redesign its own future.

In my lifetime, awareness of our planetary influence has moved from specialist knowledge into public consciousness. Now, the struggle over how to navigate that imprint plays out in full view of the cultural zeitgeist. The Information Age exposes both our brilliance and our blind spots. Globally, birth rates slip, even as the overall population swells. Economies demand growth that strains the very systems meant to sustain it. In laboratories and data centres, artificial intelligence accelerates. Meanwhile, our political and ethical reflexes lag. We carry the world's knowledge in our pockets. With it comes the power to undo the conditions of our survival. We can summon new medicines, or new pandemics; attempt to steady the climate, or push it past thresholds we barely understand.

Nonetheless, to dwell only on the prospect of collapse would disregard the more significant lesson learned from the history that brought us here. Life has survived asteroid strikes and ice ages, planetary poisonings and runaway heat. It persists by accepting change and folding it into new forms. Our own lineage exists because a catastrophe opened paths for something unexpected. Every great extinction has cleared a stage for what follows: vertebrates leaving the sea, mammals rising after the age of reptiles, our own kind emerging in the long wake of vanished species. If we are to continue, it will be by the same principles—adaptation, cooperation, imagination.

So we stand at an inflexion point, as every generation has before us; we are among the first to know in detail how improbable our existence is and the first to hold in our hands the means to end or extend it. The next chapter will be shaped in a thousand unrecorded choices—scientific, political, personal. The universe will continue, indifferent, but the continuation of awareness, of memory, and of meaning depends on what we do now.

Look back across the deep time you have just travelled with me in this book. From microbes inventing respiration in the darkness to the ecological powerhouses of rainforests and ocean waters of the present, the pattern remains one of resilience and reinvention. The same forces that forged elements in dying stars now move through your thoughts and technology. We are a late consequence of a long experiment, an animal aware of its own story, standing on the edge of futures it can scarcely imagine. Against the vast indifference of the cosmos, consciousness itself is a kind of defiance. Whether our species thrives, transforms, or vanishes, the fact remains: matter has learned to know itself. For the brief span in which that knowledge lives in us, the universe is, for a moment, hyper self-aware.

The Dusk of Humanity

And here is the most difficult thing to comprehend: your existence itself. Trace the chain backwards and you will find the odds collapsing into impossibility. For photosynthesis to spark, for eukaryotic cells to rise, for multicellularity to cohere, for each great extinction to wipe the slate and open space for the survivors to adapt—each event was a narrowing gate. Add them together, and the probability that a conscious being would emerge on this planet is indistinguishable from zero. Now fold in the contingencies of ancestry: every migration, every near-death, every birth and pairing across millions of generations. The wars and famines survived, the plagues, and the countless intimate choices—chance meetings, marriages, break-ups, first kisses that stitched each generation to the next, until your parents' lives resulted in yours.

If you were to write it as a percentage, it would begin with a decimal point, followed by zeros enough to fill every book ever printed, multiplied countless times over. Only at the end of that impossible string would a solitary 1 appear. That is your chance of being here, and yet you are. Alive and aware, against every calculation that suggests you should not, you exist.

At this juncture, the great narratives of science, history and philosophy meet. They show that we are rare, that we are endangered. Progress is real but at the same time perilous; knowledge illuminates but rarely comforts—it exposes our follies as surely as our triumphs. We have traced the mechanisms of nature only to find that the most volatile element in the equation is ourselves. Like a puppy raised by its owner to a mirror, we are tempted to look away. It reveals the adversary: our appetite for denial, our talent for greed and for hatred. But the same mirror also holds the instincts that redeem—reason, empathy, imagination. These are the means by which a species may save itself.

The window for action is open now, but it will not remain open indefinitely. Geological time and technological advancements have collided in our present; what we choose in the coming years will reverberate for epochs. It is an immense responsibility, and a strange privilege, to be alive when the stakes of life itself are laid bare.

Yet most of what shapes tomorrow will always remain beyond our command. Continents will drift, stars will flare and die, ice will advance and retreat, whether or not we are here to watch. The record of life shows that the planet's deep rhythms are indifferent to any single species. We are a brief disturbance in that continuity, and nothing we build can ensure our survival.

On the human scale, the same truth is clear. Famine endures where food could be shared, while wars are waged for ego. Rivers carry plastic to seas that will outlast us all. These acts belong to the same animal we see in ourselves, and it is easy to feel their weight as though it were our own. But no life is bound to the worst our species has wrought.

Still, the knowledge we have gathered is not useless. Understanding the history of global destruction and revival highlights the constant fragility of our survival. We cannot dictate the planet's next chapter, but we can hold off our own ending; we can give our dusk a little more room to breathe. This is the only form of stewardship that human history has allowed.

Whether our kind endures for millennia or for only a handful of centuries more, this much is ours to decide: to meet the unavoidable with clarity, to live as though the long story matters, and to leave a record of understanding that outlasts our presence. If our species must one day perish, let it be said that for a fleeting span the universe contained a species that knew what it was—and tried, for as long as it could, to keep the light of its own existence burning against the inevitable coming night.

Acknowledgements

This book draws upon a long lineage of thinkers who shaped the way I see deep time and humanity's place within it. The works of Yuval Noah Harari, Nick Lane, Tom Holland, and Richard Dawkins have each, in different ways, sharpened my understanding of evolution, history, belief, and the mechanisms that underpin life itself. Many other writers and scientists, too numerous to name individually, have contributed through their research and scholarship. My years studying geography and sustainable urban planning provided the framework through which I learned to think structurally about systems, scale, and change.

I am grateful to my parents for their steady encouragement and for cultivating an early curiosity about the natural world. Their support has been practical and unwavering, extending well beyond the writing of this book. My grandparents, aunts and uncles also played a formative role; the dinosaur toys and books they placed in my hands long ago planted questions that never quite left. My brother has been a constant source of resilience and clarity. In periods when the work felt uncertain or heavy, he offered steadiness and perspective.

The Extinct Animals Facts community has played a central role in bringing this project to life. More than half a million followers across platforms, alongside thousands of newsletter subscribers, have engaged with the themes explored here long before they took book form. The questions, debate, and sustained curiosity about deep time provided both momentum and accountability throughout the writing process.

This book was written, researched, and designed independently. Any remaining errors are mine alone.

Bibliography

Introduction

1. National Park Service. 2026. "Geologic time scale." National Park Service. Accessed February 6, 2026. https://www.nps.gov/subjects/geology/geotime.htm.
2. Encyclopaedia Britannica Editors. 2026. "Geologic time." Encyclopaedia Britannica. Accessed February 6, 2026. https://www.britannica.com/science/geologic-time.
3. Smithsonian Ocean. 2026. "The ocean throughout geologic time: An image gallery." Smithsonian Institution. Accessed February 6, 2026. https://ocean.si.edu/through-time/ancient-seas/ocean-throughout-geologic-time-image-gallery.
4. Murphy, J. Brendan, and T. H. van Andel. 2025. "Plate tectonics." Encyclopaedia Britannica. Accessed February 6, 2026. https://www.britannica.com/science/plate-tectonics.
5. Webb, Christine. 2025. "Human exceptionalism is at the root of the ecological crisis." Harvard Gazette, October 1. https://news.harvard.edu/gazette/story/2025/10/human-exceptionalism-is-at-the-root-of-the-ecological-crisis/.
6. U.S. Geological Survey. 2018. "Geologic time scale." U.S. Geological Survey. https://www.usgs.gov/media/images/geologic-time-scale-0.

Chapter 1: The Newcomers

1. O'Neill, Matthew C., et al. 2017. "Chimpanzee super strength and human skeletal muscle evolution." Proceedings of the National Academy of Sciences of the United States of America 114 (28): 6538–6543. https://doi.org/10.1073/pnas.1700190114.
2. Nielsen, Ron W. 2015. "Hyperbolic growth of the world population in the past 12,000 years." arXiv. https://doi.org/10.48550/arXiv.1510.00992.
3. Li, Xia. 2022. "The origin, spread, and impact of agriculture." Acta Anthropologica Sinica 41 (6): 1097–1108. https://doi.org/10.16359/j.1000-3193/AAS.2022.0013.
4. Bramble, Dennis M., and Daniel E. Lieberman. 2007. "The endurance running hypothesis." Journal of Human Evolution 53 (4): 434–438. https://doi.org/10.1016/j.jhevol.2007.01.012.

Chapter 2: How One Species Hijacked a Planet

1. Chure, Greg, et al. 2021. "The Anthropocene by the numbers." arXiv. https://doi.org/10.48550/arXiv.2101.09620.
2. Ewers, Robert M., and William F. Laurance. 2006. "Scale-dependent patterns of deforestation in the Brazilian Amazon." Environmental Conservation 33 (3): 203–211. https://doi.org/10.1017/S0376892906003030.
3. Ceballos, Gerardo, et al. 2015. "Accelerated modern human-induced species losses." Science Advances 1 (5): e1400253. https://doi.org/10.1126/sciadv.1400253.
4. Thompson, S. T., and P. A. O'Gorman. 2023. "Fast upper-level jet stream winds get faster under climate change." Nature Climate Change 14. https://doi.org/10.1038/s41558-023-01884-1.
5. Erb, Michael P., et al. 2022. "Reconstructing Holocene temperatures." Climate of the Past 18: 2599–2624. https://doi.org/10.5194/cp-18-2599-2022.

6. Vaughan, Adam. 2016. "Human impact has pushed Earth into the Anthropocene." The Guardian, January 7. https://www.theguardian.com/environment/2016/jan/07/human-impact-has-pushed-earth-into-the-anthropocene-scientists-say.
7. Associated Press. 2024. "Baby mammoth preserved for 50,000 years is unveiled." December 24. https://apnews.com/article/8966c2122128512ed2ec28506eb8e962.

Chapter 3: The Sixth Mass Extinction

1. Barnosky, Anthony D., et al. 2011. "Has the Earth's sixth mass extinction already arrived?" Nature 471: 51–57. https://doi.org/10.1038/nature09678.
2. Miller, Arnold I., and Matthew Lamkin. 2016. "On comparing contemporary and deep-time extinction rates." BioScience 66 (9): 785–789. https://doi.org/10.1093/biosci/biw088.
3. Ceballos, Gerardo, et al. 2015. "Accelerated modern human-induced species losses." Science Advances 1: e1400253. https://doi.org/10.1126/sciadv.1400253.
4. Sample, Ian. 2004. "Frozen Ark to save rare species." The Guardian, July 27. https://www.theguardian.com/science/2004/jul/27/biodiversity.environment.

Chapter 4: Extinctions in Recent Memory

1. Gordon, Christopher H., et al. 2015. "Canine distemper in endangered Ethiopian wolves." Emerging Infectious Diseases 21 (5). https://doi.org/10.3201/eid2105.141920.
2. IUCN–SSC Cetacean Specialist Group. 2026. "Yangtze River dolphin

(baiji)." IUCN–SSC Cetacean Specialist Group. Accessed February 4, 2026. https://iucn-csg.org/baiji/.
3. Smithsonian Magazine. 2015. "When the last of the great auks died." Smithsonian Magazine, June 9. https://www.smithsonianmag.com/smithsonian-institution/with-crush-fisherman-boot-the-last-great-auks-died-180951982/.
4. Smithsonian Magazine. 2020. "New footage of Benjamin, the last Tasmanian tiger." Smithsonian Magazine, June 1. https://www.smithsonianmag.com/smart-news/new-footage-benjamin-last-tasmanian-tiger-ever-seen-alive-180975000/.

Chapter 5: Ecology & Empire

1. Encyclopaedia Britannica Editors. 2026. "Recent extinction rates." Encyclopaedia Britannica. Accessed February 5, 2026. https://www.britannica.com/science/conservation-ecology/Recent-extinction-rates.
2. Crosby, Alfred W. 2003. The Columbian exchange: Plants, animals, and disease between the Old and New Worlds. National Humanities Center. https://nationalhumanitiescenter.org/education-material/the-columbian-exchange-plants-animals-and-disease-between-the-old-and-new-worlds/.
3. Crerar, L. D., et al. 2006. "Modelling the extinction of Steller's sea cow." Biology Letters 2 (4): 551–554. https://doi.org/10.1098/rsbl.2006.0500.
4. Linnean Society of London. 2026. "Linnaean collections." Linnean Society of London. Accessed February 5, 2026. https://www.linnean.org/research-collections/linnaean-collections.
5. New World Encyclopedia. 2026. "Georges Cuvier." New World Encyclopedia. Accessed February 5, 2026. https://www.newworldencyclopedia.org/entry/Georges_Cuvier.
6. New Zealand Birds Online. 2026. "Haast's eagle (Hieraaetus moorei)."

New Zealand Birds Online. Accessed February 5, 2026. https://www.nzbirdsonline.org.nz/species/haasts-eagle.

7. Steadman, David W., et al. 1987. "The history of avian extinction and forest conversion on Puerto Rico." Biological Conservation 39 (2): 97–111. https://doi.org/10.1016/0006-3207(87)90028-0.

Chapter 6: The Price of Bread & Circus

1. Asgari, M., et al. 2023. "Roles and factors of Caspian tiger (Panthera tigris virgata) extinction in Hyrcanian forest ecosystems." Journal of Animal Environment 15 (1): 1–10. https://doi.org/10.22034/AEJ.2023.399.
2. Black, S. A., et al. 2013. "Examining the extinction of the Barbary lion and its implications for felid conservation." PLOS ONE 8 (2): e60174. https://doi.org/10.1371/journal.pone.0060174.
3. Crosby, Alfred W. 2003. The Columbian exchange. National Humanities Center. https://nationalhumanitiescenter.org/education-material/the-columbian-exchange-plants-animals-and-disease-between-the-old-and-new-worlds/.
4. Crerar, L. D., et al. 2006. "Modelling the extinction of Steller's sea cow." Biology Letters 2 (4): 551–554. https://doi.org/10.1098/rsbl.2006.0500.
5. Epplett Jr., William C. 2002. Animal spectacula of the Roman Empire. Master's thesis, University of British Columbia.
6. Lindberg, Nicholas. 2019. "The emperor and his animals: The acquisition of exotic beasts for imperial venationes." Greece & Rome 66 (2): 251–263. https://doi.org/10.1017/S0017383519000068.
7. Metropolitan Museum of Art. 2026. "Hippopotami in ancient Egypt." Metropolitan Museum of Art. Accessed February 5, 2026. https://www.metmuseum.org/essays/hippopotami-in-ancient-egypt.
8. Metropolitan Museum of Art. 2026. "Roman games: Playing with

animals." Metropolitan Museum of Art. Accessed February 5, 2026. https://www.metmuseum.org/essays/roman-games-playing-with-animals.

9. Schwimer, L., et al. 2024. "Petroglyphs as a paleobiogeographic guide for Arabian ostrich ecology and reintroduction." Journal of Arid Environments 224: 105222. https://doi.org/10.1016/j.jaridenv.2024.105222.
10. Tsahar, E., et al. 2009. "Distribution and extinction of ungulates during the Holocene of the southern Levant." PLOS ONE 4 (4): e5316. https://doi.org/10.1371/journal.pone.0005316.
11. United Nations Environment Programme. 2026. The Mesopotamian marshlands: A human and environmental catastrophe. Accessed February 5, 2026. https://eoimages.gsfc.nasa.gov/images/imagerecords/1000/1716/meso2.pdf.
12. Yong, Ed. 2008. "Elephants recognise themselves in a mirror." National Geographic, September 28. https://www.nationalgeographic.com/science/article/elephants-recognise-themselves-in-mirror.

Chapter 7: The Weight of Innovation

1. Atwood, Roger. 2011. "The Nok of Nigeria." Archaeology Magazine, July 1. https://www.archaeology.org/issues/65-1107/features/2647-nok-of-nigeria.
2. Butzer, Karl W. 2012. "Collapse, environment, and society." Proceedings of the National Academy of Sciences of the United States of America 109 (10): 3632–3639. https://doi.org/10.1073/pnas.1114845109.
3. Cline, Eric H. 2014. 1177 B.C.: The year civilization collapsed. Princeton, NJ: Princeton University Press.
4. Mark, Joshua J. 2026. "Greek Dark Age." World History Encyclopedia. Accessed February 5, 2026. https://www.worldhistory.org/Greek_Dark_Age/.

5. Matthiesen, Henrik, et al. 2022. "Wetland archaeology and the impact of climate change." Antiquity 96 (390). https://doi.org/10.15184/aqy.2022.112.
6. Roberts, Benjamin W., and Christopher P. Thornton, eds. 2014. Archaeometallurgy in global perspective. New York: Springer.
7. Schmandt-Besserat, Denise. 1996. How writing came about. Austin: University of Texas Press.
8. Smith, Roff. 2013. "Drought led to collapse of civilizations, study says." National Geographic, October 25. https://www.nationalgeographic.com/culture/article/131024-drought-bronze-age-pollen-archaeology.

Chapter 8: Neither Beast nor God

1. Barzun, Jacques. 2025. "The metal ages." In History of Europe. Encyclopaedia Britannica. Accessed February 5, 2026. https://www.britannica.com/topic/history-of-Europe/The-chronology-of-the-Metal-Ages.
2. Bocquet-Appel, Jean-Pierre. 2002. "Paleoanthropological traces of a Neolithic demographic transition." Current Anthropology 43 (4): 637–650. https://doi.org/10.1086/342429.
3. Guilmartin, John F. 2026. "Walls of Jericho." Encyclopaedia Britannica. Accessed February 5, 2026. https://www.britannica.com/place/walls-of-Jericho.
4. Hoffecker, John F., et al. 2016. "Beringia and the global dispersal of modern humans." Evolutionary Anthropology 25 (2): 64–78. https://doi.org/10.1002/evan.21478.
5. Masood, Farah. 2020. "The effect of agriculture on health in Neolithic populations in the Levant." Pathways: A Graduate Journal of Anthropology 1 (1). https://doi.org/10.29173/pathways11.
6. Parker, Chris, et al. 2023. "Violence-related trauma in Neolithic Europe indicates early warfare." Proceedings of the National Academy of

Sciences of the United States of America 120 (4). https://doi.org/10.1073/pnas.2214720120.

7. Smith, Michael E., et al. 2025. "The deep roots of inequality in human societies." Proceedings of the National Academy of Sciences of the United States of America 122 (16). https://doi.org/10.1073/pnas.2400691122.
8. Whittle, Alasdair. 1996. Europe in the Neolithic: The creation of new worlds. Cambridge: Cambridge University Press.
9. Wymer, John. 1982. The Mesolithic settlement of northern Europe. London: Routledge.

Chapter 9: The Planet with Frostbite

1. Bamber, Jonathan L., and Robert L. England. 1996. "Sea level changes, continental shelf morphology, and global paleoecological constraints." Palaeogeography, Palaeoclimatology, Palaeoecology 121 (3–4): 259–271. https://doi.org/10.1016/0031-0182(95)00085-2.
2. Covey, Curt. 1984. "Reconstructing Pleistocene climatic change from oxygen isotopes." Annals of Glaciology 5: 43–46. https://doi.org/10.3189/1984AoG5-1-43-46.
3. Graversen, Rune G., et al. 2024. "Abrupt increase in Greenland melt enhanced by atmospheric wave changes." Climate Dynamics 62. https://doi.org/10.1007/s00382-024-07271-6.
4. Hu, Haipeng, et al. 2023. "Genomic inference of a severe human bottleneck." Science 381 (6657): 979–984. https://doi.org/10.1126/science.abq7487.
5. NASA Science. 2026. "Milankovitch cycles and their role in Earth's climate." NASA. Accessed February 6, 2026. https://science.nasa.gov/science-research/earth-science/milankovitch-orbital-cycles-and-their-role-in-earths-climate.
6. Taylor, Kendrick C., et al. 1997. "The Holocene–Younger Dryas

transition." Science 278 (5339): 825–827. https://doi.org/10.1126/science.278.5339.825.

7. U.S. Geological Survey. 2026. "Glaciers extended over much of Europe during the last Ice Age." U.S. Geological Survey. Accessed February 6, 2026. https://www.usgs.gov/media/images/glaciers-extended-over-much-europe-during-last-ice-age.

Chapter 10: Life on the Mammoth Steppe

1. Elias, Scott A. 2015. "Beringia: Lost world of the Ice Age." Alaska Park Science. National Park Service. Accessed February 6, 2026. https://www.nps.gov/articles/aps-v12-i2-c8.htm.
2. Zimov, Sergey A., et al. 2012. "Mammoth steppe: A high-productivity phenomenon." Quaternary Science Reviews 57: 26–45. https://doi.org/10.1016/j.quascirev.2012.10.005.
3. Gould, Stephen Jay. 1974. "The origin and function of 'bizarre' structures." Evolution 28 (2): 191–220. https://doi.org/10.1111/j.1558-5646.1974.tb00740.x.
4. Faith, J. Tyler. 2011. "Late Pleistocene climate change and megafaunal extinctions." Quaternary Science Reviews 30 (13–14): 1675–1680. https://doi.org/10.1016/j.quascirev.2011.03.011.
5. Dehasque, Marianne, et al. 2024. "Temporal dynamics of woolly mammoth genome erosion." Cell 187 (6): 1549–1564. https://www.cell.com/cell/fulltext/S0092-8674(24)00577-4.
6. Devlin, Hannah. 2024. "Freak event probably killed last woolly mammoths." The Guardian, June 27. https://www.theguardian.com/science/article/2024/jun/27/last-woolly-mammoths-arctic-island.

Chapter 11: The Paleo-Economics of Collapse

1. Alroy, John. 2001. "A multispecies overkill simulation of the end-Pleistocene megafaunal mass extinction." Science 292 (5523): 1893–1896. https://doi.org/10.1126/science.1059342.
2. Ben-Dor, Miki, and Ran Barkai. 2024. "A matter of fat: Hunting preferences affected Pleistocene megafaunal extinctions and human evolution." Quaternary Science Reviews 331: 108660. https://doi.org/10.1016/j.quascirev.2024.108660.
3. Bird, Rebecca J., et al. 2008. "The 'fire-stick farming' hypothesis: Australian Aboriginal foraging strategies, biodiversity, and anthropogenic fire mosaics." Proceedings of the National Academy of Sciences of the United States of America 105 (39): 14796–14801. https://doi.org/10.1073/pnas.0804757105.
4. Cuffey, Kurt M., et al. 1995. "Large Arctic temperature change at the Wisconsin–Holocene glacial transition." Science 270 (5235): 455–458. https://doi.org/10.1126/science.270.5235.455.
5. Faith, J. Tyler. 2011. "Late Pleistocene climate change, nutrient cycling, and the megafaunal extinctions in North America." Quaternary Science Reviews 30 (13–14): 1675–1680. https://doi.org/10.1016/j.quascirev.2011.03.011.
6. Firestone, Richard B., et al. 2007. "Evidence for an extraterrestrial impact 12,900 years ago that contributed to the megafaunal extinctions and the Younger Dryas cooling." Proceedings of the National Academy of Sciences of the United States of America 104 (41): 16016–16021. https://doi.org/10.1073/pnas.0706977104.
7. Holliday, Vance T., et al. 2011. "The Younger Dryas impact hypothesis: A requiem." Earth-Science Reviews 106 (1–2): 1–30.
8. Johnson, William H. 2026. "Pleistocene Epoch." Encyclopaedia Britannica. Accessed February 6, 2026. https://www.britannica.com/science/Pleistocene-Epoch.
9. Lorenzen, Eline D., et al. 2011. "Species-specific responses of late Quaternary megafauna to climate and humans." Proceedings of the National Academy of Sciences of the United States of America 108 (33): 13409–13413. https://doi.org/10.1073/pnas.1102974108.

10. Monteath, Alistair J., et al. 2023. "Relict permafrost preserves megafauna, insects, pollen, soils and pore-ice isotopes of the mammoth steppe and its collapse in central Yukon." Quaternary Science Reviews 299: 107878. https://doi.org/10.1016/j.quascirev.2022.107878.
11. Politis, Gustavo G., et al. 2019. "Campo Laborde: A late Pleistocene giant ground sloth kill and butchery site in the Pampas (Argentina)." Science Advances 5 (3): eaau4546. https://doi.org/10.1126/sciadv.aau4546.
12. Sandom, Christopher, Søren Faurby, and Jens-Christian Svenning. 2014. "Global late Quaternary megafauna extinctions linked to humans, not climate change." Proceedings of the Royal Society B 281 (1787): 20140860. https://doi.org/10.1098/rspb.2014.0860.
13. Taylor, Kendrick C., et al. 1997. "The Holocene–Younger Dryas transition recorded at Summit, Greenland." Science 278 (5339): 825–827. https://doi.org/10.1126/science.278.5339.825.

Chapter 12: *Primordium Mentis*

1. Alemseged, Zeresenay. 2014. "Early hominins." In The Cambridge world prehistory, edited by Colin Renfrew and Paul Bahn. Cambridge: Cambridge University Press.
2. Ben-Dor, Miki, and Ran Barkai. 2024. "A matter of fat: Hunting preferences affected Pleistocene megafaunal extinctions and human evolution." Quaternary Science Reviews 331: 108660. https://doi.org/10.1016/j.quascirev.2024.108660.
3. Bird, Rebecca J., et al. 2008. "The 'fire-stick farming' hypothesis." Proceedings of the National Academy of Sciences of the United States of America 105 (39): 14796–14801. https://doi.org/10.1073/pnas.0804757105.
4. Burtsev, Mikhail, Konstantin Anokhin, and Patrick Bateson. 2024. "Facilitation of evolution by plasticity scales with phenotypic complexity."

Animals 14 (19): 2804. https://doi.org/10.3390/ani14192804.

5. Drewnowski, Adam. 2012. "Sweetness and food preference." Journal of Nutrition 142 (6): 1142S–1148S. https://doi.org/10.3945/jn.111.149575.
6. Huerta-Sánchez, Emilia, et al. 2014. "Altitude adaptation in Tibetans caused by introgression of Denisovan-like DNA." Nature 512 (7513): 194–197. https://doi.org/10.1038/nature13408.
7. Lordkipanidze, David, et al. 2005. "The earliest toothless hominin skull." Nature 434 (7034): 717–718. https://doi.org/10.1038/nature03496.
8. Pääbo, Svante, et al. 2010. "Genetic history of an archaic hominin group from Denisova Cave in Siberia." Nature 468 (7327): 1053–1060. https://doi.org/10.1038/nature09710.
9. Potts, Richard. 1998. "Environmental hypotheses of hominin evolution." Yearbook of Physical Anthropology 41: 93–136.
10. Reich, David, et al. 2011. "Denisova admixture and the first modern human dispersals into Southeast Asia and Oceania." Cell 146 (3): 279–287. https://doi.org/10.1016/j.cell.2011.06.029.
11. Smithsonian Institution. 2026. "Homo neanderthalensis." Human Origins Program. Accessed February 6, 2026. https://humanorigins.si.edu.
12. Smithsonian Institution. 2026. "Homo erectus." Human Origins Program. Accessed February 6, 2026. https://humanorigins.si.edu.
13. Smithsonian Institution. 2026. "Australopithecus afarensis." Human Origins Program. Accessed February 6, 2026. https://humanorigins.si.edu.
14. Smithsonian Institution. 2026. "AL 288-1 (Lucy)." Human Origins Program. Accessed February 6, 2026. https://humanorigins.si.edu.

Chapter 13: Continents Adrift

1. Angst, Delphine, et al. 2013. "Terror birds (Phorusrhacidae) from

the Eocene of Europe imply transoceanic dispersal." PLOS ONE 8 (11): e80357. https://doi.org/10.1371/journal.pone.0080357.

2. Bocherens, Hervé, et al. 2017. "Isotopic insight on paleodiet of extinct Pleistocene megafaunal xenarthrans from Argentina." Gondwana Research 48: 7–14. https://doi.org/10.1016/j.gr.2017.03.002.
3. Domingo, Laura, et al. 2020. "The Great American Biotic Interchange revisited." Scientific Reports 10: 1175. https://doi.org/10.1038/s41598-020-58093-1.
4. Webb, S. David. 2006. "The Great American Biotic Interchange: Patterns and processes." Annals of the Missouri Botanical Garden 93 (2): 245–257.
5. Welker, Frido, et al. 2015. "Ancient proteins resolve the evolutionary history of Darwin's South American ungulates." Nature 522: 81–84. https://doi.org/10.1038/nature14249.
6. Wroe, Stephen, et al. 2007. "Cranial mechanics compared in extinct marsupial and extant African lions." Journal of Zoology 274 (4): 332–339. https://doi.org/10.1111/j.1469-7998.2007.00302.x.
7. Woodburne, Michael O. 2010. "The Great American Biotic Interchange." Journal of Mammalian Evolution 17: 245–264. https://doi.org/10.1007/s10914-010-9146-8.
8. Australian Museum. 2026. "Diprotodon optatum." Australian Museum. Accessed February 5, 2026. https://australian.museum.

Chapter 14: Blueprints of the Familiar

1. Schneider, Birgit, and Andreas Schmittner. 2006. "Simulating the impact of the Panamanian Seaway closure on ocean circulation." Earth and Planetary Science Letters 246 (3–4): 367–380. https://doi.org/10.1016/j.epsl.2006.04.028.
2. Christin, Pascal-Antoine, et al. 2008. "Oligocene CO_2 decline promoted C_4 photosynthesis in grasses." Current Biology 18 (1): 37–43. https://d

oi.org/10.1016/j.cub.2007.11.058.

3. Edwards, Erika J., et al. 2010. "The origins of C_4 grasslands." Science 328 (5978): 587–591. https://doi.org/10.1126/science.1185009.
4. Clay, Keith, et al. 2005. "Herbivores cause a rapid increase in hereditary symbiosis." Proceedings of the National Academy of Sciences of the United States of America 102 (35): 12465–12470. https://doi.org/10.1073/pnas.0501763102.
5. Li, Ziye, et al. 2023. "Neogene burial of organic carbon in the global ocean." Nature 613: 90–95. https://doi.org/10.1038/s41586-022-05594-6.
6. Rule, James P., et al. 2023. "Giant baleen whales emerged from a cold southern cradle." Proceedings of the Royal Society B 290: 20232177. https://doi.org/10.1098/rspb.2023.2177.
7. Encyclopaedia Britannica Editors. 2026. "Indricotherium." Encyclopaedia Britannica. Accessed February 2026. https://www.britannica.com/animal/Indricotherium.

Chapter 15: The Inheritors of Hell

1. Kemp, T. S. 2005. The origin and evolution of mammals. Oxford: Oxford University Press.
2. Encyclopaedia Britannica Editors. 2025. "Chicxulub Crater." Encyclopaedia Britannica. Accessed February 5, 2026. https://www.britannica.com/place/Chicxulub.
3. NASA Science. 2026. "Deep Impact and the mass extinction of species 65 million years ago." NASA. Accessed February 5, 2026. https://science.nasa.gov.
4. Longrich, Nicholas R., et al. 2016. "Severe extinction and rapid recovery of mammals across the Cretaceous–Palaeogene boundary." Journal of Evolutionary Biology 29 (8): 1495–1512.
5. Vajda, Vivi, et al. 2001. "Indication of global deforestation at the

Cretaceous–Tertiary boundary." Science 294 (5547): 1700–1702.
6. Smithsonian Institution. 2026. "Pantolambda bathmodon Cope." Smithsonian Institution. Accessed February 5, 2026. https://www.si.edu.
7. Christison, Brigid E., et al. 2022. "Dietary niches of creodonts and carnivorans." Journal of Mammalogy 103 (1): 2–17.
8. Chester, Stephen G. B., et al. 2015. "Oldest known euarchontan tarsals." Proceedings of the National Academy of Sciences of the United States of America 112 (3): 803–808.

Chapter 16: Dragons in the Dark

1. Encyclopaedia Britannica Editors. 2026. "How long did dinosaurs live?" Encyclopaedia Britannica. Accessed February 5, 2026. https://www.britannica.com/science/How-Long-Did-Dinosaurs-Live.
2. Encyclopaedia Britannica Editors. 2025. "Permian–Triassic extinctions." In Triassic Period. Encyclopaedia Britannica. Accessed February 5, 2026. https://www.britannica.com/science/Triassic-Period/Permian-Triassic-extinctions.
3. Dececchi, Thomas A., et al. 2024. "Theropod trackways as indirect evidence of pre-avian aerial behavior." Proceedings of the National Academy of Sciences of the United States of America 121 (44). https://doi.org/10.1073/pnas.2413810121.
4. Geggel, Laura. 2017. "T. rex could pulverize bones with a force of nearly 8,000 pounds." Live Science, May 17. https://www.livescience.com/59134-tyrannosaurus-rex-bite-could-pulverize-bones.html.
5. Logan, Amanda. 2025. "Triassic Period." Encyclopaedia Britannica. Accessed February 5, 2026. https://www.britannica.com/science/Triassic-Period.
6. NASA Astrobiology Institute. 2001. "Do we know what killed the dinosaurs?" NASA Astrobiology Institute, June 15. https://astrobiology.nasa.gov/nai/articles/2001/6/15/do-we-know-what-killed-the-

dinosaurs/.

7. Oxford University Museum of Natural History. 2026. "The Oxfordshire dinosaurs." Oxford University Museum of Natural History. Accessed February 5, 2026. https://oumnh.ox.ac.uk/learn-oxfordshire-dinosaurs.
8. Switek, Brian. 2014. "Five things we don't know about Tyrannosaurus rex." Smithsonian Magazine, April 15. https://www.smithsonianmag.com/science-nature/five-things-we-dont-know-about-tyrannosaurus-rex-180951072/.

Chapter 17: Thrones of Tyrants, Crowns of Horns

1. Bazzi, Mohamed, et al. 2021. "Tooth morphology elucidates shark evolution across the end-Cretaceous mass extinction." PLOS Biology 19 (8): e3001108. https://doi.org/10.1371/journal.pbio.3001108.
2. Brinkhuis, Henk, et al. 2008. "Long-term sea-level fluctuations driven by ocean basin dynamics." Science 321 (5887): 1351–1354. https://doi.org/10.1126/science.1156909.
3. Encyclopaedia Britannica Editors. 2025. "Cretaceous Period." Encyclopaedia Britannica. Accessed February 6, 2026. https://www.britannica.com/science/Cretaceous-Period.
4. Encyclopaedia Britannica Editors. 2023. "Quetzalcoatlus." Encyclopaedia Britannica. Accessed February 6, 2026. https://www.britannica.com/animal/Quetzalcoatlus.
5. Encyclopaedia Britannica Editors. 2026. "Tyrannosaurus rex." Encyclopaedia Britannica. Accessed February 6, 2026. https://www.britannica.com/animal/Tyrannosaurus-rex.
6. Dececchi, Thomas A., et al. 2024. "Theropod trackways as indirect evidence of pre-avian aerial behavior." Proceedings of the National Academy of Sciences of the United States of America 121 (44). https://doi.org/10.1073/pnas.2413810121.

7. Farke, Andrew A., et al. 2009. "Evidence of combat in Triceratops." PLOS ONE 4 (1): e4252. https://doi.org/10.1371/journal.pone.0004252.
8. Fiorillo, Anthony R., et al. 2018. "An unusual association of hadrosaur and therizinosaur tracks within Late Cretaceous rocks of Denali National Park, Alaska." Scientific Reports 8: 12395. https://doi.org/10.1038/s41598-018-30110-8.
9. Habib, Michael B. 2019. "Pterosaurs were monsters of the Mesozoic skies." Scientific American 321 (4): 58–65. https://doi.org/10.1038/scientificamerican1019-58.
10. Konishi, Takuya, et al. 2014. "A mosasaur's last meal." National Geographic, November 12. https://www.nationalgeographic.com/science/article/a-mosasaurs-last-meal.
11. Rothman, Daniel H. 2001. "The mid-Cretaceous superplume, carbon dioxide, and global warming." Earth and Planetary Science Letters 193 (3–4): 337–344. https://doi.org/10.1016/S0012-821X(01)00404-0.
12. Spicer, Robert A., et al. 1987. "Alaskan Cretaceous–Tertiary floras and Arctic origins." Paleobiology 13 (1): 1–13. https://doi.org/10.1017/S0094837300008599.
13. Vajda, Vivi, et al. 2001. "Indication of global deforestation at the Cretaceous–Tertiary boundary by New Zealand fern spike." Science 294 (5547): 1700–1702. https://doi.org/10.1126/science.1064706.

Chapter 18: The Ascent of Giants

1. Bedard, Jean. 1988. "Triassic–Jurassic rifting and opening of the Atlantic: An overview." In Developments in Geotectonics, vol. 22, 41–79. Amsterdam: Elsevier. https://doi.org/10.1016/B978-0-444-42903-2.50008-7.
2. Daniels, Christopher B., and Jonathan P. Pratt. 1992. "Breathing in long-necked dinosaurs: Did the sauropods have bird lungs?" Comparative Biochemistry and Physiology Part A 101 (1): 43–46. https://doi.or

g/10.1016/0300-9629(92)90625-Z.

3. D'Emic, Michael D. 2011. "Sauropod dinosaurs: The evolution of gigantism." In Biology of the sauropod dinosaurs: Understanding the life of giants, edited by P. Martin Sander et al. Bloomington: Indiana University Press.
4. Lubick, Naomi. 2018. "Making Jurassic tracks in the Jura." Earth Magazine, June 12. Accessed February 5, 2026. https://www.earthmagazine.org/article/making-jurassic-tracks-jura.
5. Natural History Museum (London). 2026. "Brachiosaurus." Natural History Museum. Accessed February 5, 2026. https://www.nhm.ac.uk/discover/dino-directory/brachiosaurus.html.
6. National Geographic. 2026. "Jurassic Period." National Geographic. Accessed February 5, 2026. https://www.nationalgeographic.com/science/article/jurassic.
7. PBS. 2026. "Evolution: Change: Deep time." PBS. Accessed February 5, 2026. https://www.pbs.org/wgbh/evolution/change/deeptime/jurassic.html.
8. Ruxton, Graeme D., Gordon F. Birchard, and D. Charles Deeming. 2014. "Incubation time as an important influence on egg production and distribution into clutches for sauropod dinosaurs." Paleobiology 40 (3): 323–330. https://doi.org/10.1666/13004.
9. Sander, P. Martin, et al. 2011. "Biology of the sauropod dinosaurs: The evolution of gigantism." Biological Reviews 86 (1): 117–155. https://doi.org/10.1111/j.1469-185X.2010.00137.x.
10. Wedel, Mathew J. 2003. "Vertebral pneumaticity, air sacs, and the physiology of sauropod dinosaurs." Paleobiology 29 (2): 243–255.

Chapter 19: Interregnum of the Archosaurs

1. Bronzati, Marcos, et al. 2015. "Macroevolutionary dynamics of crocodylomorph body size evolution." Current Biology 25 (22): 1–7.

https://doi.org/10.1016/j.cub.2015.08.021.

2. Brusatte, Stephen L., et al. 2011. "Macroevolutionary patterns in the evolutionary radiation of archosaurs." Earth and Environmental Science Transactions of the Royal Society of Edinburgh 101: 367–382. https://doi.org/10.1017/S1755691011000130.
3. Butler, Richard J., et al. 2012. "Reassessment of the evidence for postcranial skeletal pneumaticity in Triassic archosaurs." PLOS ONE 7 (3): e34094. https://doi.org/10.1371/journal.pone.0034094.
4. Nesbitt, Sterling J. 2011. The early evolution of archosaurs. Bulletin of the American Museum of Natural History 352: 1–292. https://doi.org/10.1206/352.1.
5. Petersen, Henrik I., and Sofie Lindström. 2012. "Synchronous wildfire activity rise and mire deforestation at the Triassic–Jurassic boundary." PLOS ONE 7 (10): e47236. https://doi.org/10.1371/journal.pone.0047236.
6. Ruhl, Micha, et al. 2005. "Rising atmospheric CO_2 as a possible trigger for the end-Triassic mass extinction." Palaeogeography, Palaeoclimatology, Palaeoecology 217 (3–4): 223–242. https://doi.org/10.1016/j.palaeo.2004.12.004.
7. Twitchett, Richard J., et al. 2006. "Rapid vertebrate recuperation in the Karoo Basin." Journal of African Earth Sciences 45 (4–5): 502–514. https://doi.org/10.1016/j.jafrearsci.2006.04.006.
8. Viglietti, Pia A., et al. 2022. "The paleobiology and paleoecology of South African Lystrosaurus." Palaeontologia Africana 57: 1–15. https://doi.org/10.25100/pa.v57i0.10000.
9. Whitney, Meredith R., and Christian A. Sidor. 2020. "Evidence of torpor in the tusks of Lystrosaurus." Communications Biology 3: 471. https://doi.org/10.1038/s42003-020-01207-6.
10. National Park Service. 2026. "Triassic Period—251.9 to 201.3 MYA." National Park Service. Accessed February 5, 2026. https://www.nps.gov/articles/000/triassic-period.htm.

Chapter 20: The Anatomy of Planetary Failure

1. Benton, Michael J., and Richard J. Twitchett. 2003. "How to kill (almost) all life: The end-Permian extinction event." Trends in Ecology & Evolution 18 (7): 358–365.
2. Burgess, Seth D., and Benjamin A. Black. 2025. "The anatomy and lethality of the Siberian Traps large igneous province." Annual Review of Earth and Planetary Sciences 53: 567–594. https://doi.org/10.1146/annurev-earth-040722-105544.
3. Burgess, Seth D., Benjamin A. Black, and Samuel A. Bowring. 2017. "Subsurface magma triggers Earth's most severe extinction." U.S. Geological Survey, July 31. Accessed February 5, 2026. https://www.usgs.gov/news/national-news-release/subsurface-magma-triggers-earths-most-severe-extinction.
4. Burgess, Seth D., Samuel A. Bowring, and Shuzhong Shen. 2014. "High-precision timeline for Earth's most severe extinction." Proceedings of the National Academy of Sciences of the United States of America 111. https://doi.org/10.1073/pnas.1317692111.
5. Dal Corso, Jacopo, et al. 2022. "Environmental crises at the Permian–Triassic mass extinction." Nature Reviews Earth & Environment 3: 197–214. https://doi.org/10.1038/s43017-021-00259-4.
6. Ivanov, Andrey V., et al. 2013. "Siberian Traps large igneous province." Lithos 178: 1–15. https://doi.org/10.1016/j.lithos.2013.01.001.
7. Encyclopaedia Britannica. 2026. "Permian extinction." Encyclopaedia Britannica. Accessed February 5, 2026. https://www.britannica.com/science/Permian-extinction.

Chapter 21: The Bedrock of Industrial Might

1. National Park Service. 2026. "Mississippian Period (358.9 to 323.2

million years ago)." National Park Service. Accessed February 6, 2026. https://www.nps.gov/articles/000/mississippian-period.htm.

2. National Park Service. 2026. "Pennsylvanian Period (323.2 to 298.9 million years ago)." National Park Service. Accessed February 6, 2026. https://www.nps.gov/articles/000/pennsylvanian-period.htm.
3. National Geographic Society. 2023. "Carboniferous Period." National Geographic. Accessed February 6, 2026. https://www.nationalgeographic.com/science/article/carboniferous.
4. Berner, Robert A. 1998. "The carbon cycle and carbon dioxide over Phanerozoic time: The role of land plants." Philosophical Transactions of the Royal Socicty B 353 (1365): 75–82. https://royalsocietypublishing.org/doi/10.1098/rstb.1998.0192.
5. Nelsen, Matthew P., et al. 2016. "Delayed fungal evolution did not cause the Paleozoic peak in coal production." Proceedings of the National Academy of Sciences of the United States of America 113 (9): 2442–2447. https://www.pnas.org/doi/10.1073/pnas.1517943113.
6. Glasspool, Ian J., and Andrew C. Scott. 2010. "Phanerozoic concentrations of atmospheric oxygen reconstructed from sedimentary charcoal." Nature Geoscience 3: 627–630. https://www.nature.com/articles/ngeo923.
7. Belcher, Claire M., et al. 2008. "Fire sensitivity to low atmospheric oxygen concentration." Science 321 (5885): 1647–1650. https://www.science.org/doi/10.1126/science.1160017.
8. Royal Holloway, University of London. 2010. "Oxygen fuels the fires of time." ScienceDaily, August 2. Accessed February 6, 2026. https://www.sciencedaily.com/releases/2010/08/100802091125.htm.
9. Benton, Michael J., et al. 2017. "Diversity changes in Palaeozoic terrestrial vertebrates." Proceedings of the Royal Society B 284 (1867). https://royalsocietypublishing.org/doi/10.1098/rspb.2017.2730.
10. Wheatley, Patrick J., et al. 2018. "Palaeozoic giant dragonflies were hawker predators." Scientific Reports 8: 12149. https://www.nature.com/articles/s41598-018-30629-w.
11. Briggs, Derek E. G., et al. 1975. "The arthropleurids from the Carbonif-

erous of North America." Journal of Paleontology 49 (3): 484–495. https://www.jstor.org/stable/1303372.

12. Clapham, Matthew E., and Jered J. Karr. 2012. "Environmental and biotic controls on insect body size." Proceedings of the National Academy of Sciences of the United States of America 109 (27): 10927–10930. https://www.pnas.org/doi/10.1073/pnas.1204026109.
13. Sahney, Sarda, Michael J. Benton, and Howard J. Falcon-Lang. 2010. "Rainforest collapse triggered Pennsylvanian tetrapod diversification in Euramerica." Geology 38 (12): 1079–1082. https://pubs.geoscienceworld.org/gsa/geology/article/38/12/1079/130264.
14. Clack, Jennifer A. 2012. Gaining ground: The origin and evolution of tetrapods. 2nd ed. Bloomington: Indiana University Press.
15. Encyclopaedia Britannica. 2026. "Amniote." Encyclopaedia Britannica. Accessed February 6, 2026. https://www.britannica.com/animal/amniote.
16. Smil, Vaclav. 2017. Energy and civilization: A history. Cambridge, MA: MIT Press.
17. U.S. Energy Information Administration. 2026. "Coal explained." U.S. Energy Information Administration. Accessed February 6, 2026. https://www.eia.gov/energyexplained/coal/.

Chapter 22: Claiming *Terra Incognito*

1. Encyclopaedia Britannica. 2026. "Archaeopteris." Encyclopaedia Britannica. Accessed February 6, 2026. https://www.britannica.com/plant/Archaeopteris.
2. Encyclopaedia Britannica. 2024. "Devonian extinctions." Encyclopaedia Britannica. Accessed February 6, 2026. https://www.britannica.com/science/Devonian-extinctions.
3. Encyclopaedia Britannica. 2026. "Insect fossil record." Encyclopaedia

Britannica. Accessed February 6, 2026. https://www.britannica.com/animal/insect/Insect-fossil-record.

4. Encyclopaedia Britannica. 2026. "Plant: Evolution and paleobotany." Encyclopaedia Britannica. Accessed February 6, 2026. https://www.britannica.com/plant/plant/Evolution-and-paleobotany.
5. Chapman, R. F. 2013. The insects: Structure and function. 5th ed. Cambridge: Cambridge University Press.
6. Copper, Paul. 2011. "100 million years of reef prosperity and collapse: Ordovician to Devonian interval." The Paleontological Society Papers 17: 15–32. https://doi.org/10.1017/S1089332600002424.
7. Daeschler, Edward B., Neil H. Shubin, and Farish A. Jenkins Jr. 2006. "A Devonian tetrapod-like fish and the evolution of the tetrapod body plan." Nature 440 (7085): 757–763. https://doi.org/10.1038/nature04639.
8. Field Museum of Natural History. 2026. "Devonian stromatoporoid and tabulate coral reefs." Field Museum of Natural History. Accessed February 6, 2026. https://silurian-reef.fieldmuseum.org/narrative/483.
9. Gensel, Patricia G., and Dianne Edwards. 2001. Plants invade the land: Evolutionary and environmental perspectives. New York: Columbia University Press.
10. Kenrick, Paul, and Peter R. Crane. 1997. The origin and early diversification of land plants: A cladistic study. Washington, DC: Smithsonian Institution Press.

Chapter 23: Webs Beneath the Waves

1. Holland, Steven M. 2026. "Ordovician period: Oceanography." Encyclopaedia Britannica. Accessed February 6, 2026. https://www.britannica.com/science/Ordovician-Period/Oceanography.
2. Kozik, Nevin P., et al. 2022. "Rapid marine oxygen variability: Driver

of the Late Ordovician mass extinction." Science Advances 8 (46): eabn8345. https://doi.org/10.1126/sciadv.abn8345.
3. Zhang, Lin-Na, et al. 2025. "Biogeographic dynamics of graptolites during the Late Ordovician mass extinction in South China." Palaeoworld 34 (2): 100868. https://doi.org/10.1016/j.palwor.2024.06.008.
4. Krug, A. Z., and M. E. Patzkowsky. 2004. "Rapid recovery from the Late Ordovician mass extinction." Proceedings of the National Academy of Sciences of the United States of America 101 (49): 17006–17011. https://doi.org/10.1073/pnas.0405199102.
5. Huang, Bing, et al. 2024. "A new post-LOME (Late Ordovician mass extinction) recovery brachiopod fauna from South China." Journal of Paleontology 98 (3): 366–377. https://doi.org/10.1017/jpa.2024.14.

Chapter 24: The Bakuhatsu Upheaval

1. Fox, Douglas. 2016. "What sparked the Cambrian explosion?" Scientific American, February 16. https://www.scientificamerican.com/article/what-sparked-the-cambrian-explosion1/.
2. Akam, Michael. 1995. "Hox genes and the evolution of diverse body plans." Development 121 (1): 1–8. https://doi.org/10.1242/dev.121.1.1.
3. Encyclopedia of the Environment. n.d. "The first complex ecosystems." Encyclopedia of the Environment. Accessed February 6, 2026. https://www.encyclopedie-environnement.org/en/life/first-complex-ecosystems-2/.
4. Murdock, Duncan J. E. 2020. "The biomineralization toolkit and the origin of animal skeletons." Biological Reviews 95 (5): 1372–1392. https://doi.org/10.1111/brv.12614.
5. Flannery, Timothy Fridtjof. n.d. "Cambrian explosion." Encyclopaedia Britannica. Accessed February 6, 2026. https://www.britannica.com/science/Cambrian-explosion.
6. British Geological Survey. n.d. "Trilobites." British Geological Survey.

Accessed February 6, 2026. https://www.bgs.ac.uk/discovering-geology/fossils-and-geological-time/trilobites/.

7. Lamb, Trevor D. 2011. "Evolution of the eye." Scientific American, July 1. https://www.scientificamerican.com/article/evolution-of-the-eye/.
8. Clarkson, Euan N. K., Riccardo Levi-Setti, and Gábor Horváth. 2006. "The eyes of trilobites: The oldest preserved visual system." Arthropod Structure & Development 35 (4): 247–259. https://doi.org/10.1016/j.asd.2006.08.002.
9. Erwin, Douglas H., and James W. Valentine. 2013. The Cambrian explosion: The construction of animal biodiversity. Greenwood Village, CO: Roberts & Company Publishers.
10. Peterson, Kevin J. 2008. "Molecular paleobiology and the Cambrian explosion: 21st century answers to 19th century problems." The Paleontological Society Papers 14: 105–116. https://doi.org/10.1017/S1089332600001637.
11. International Commission on Stratigraphy. n.d. "International chronostratigraphic chart." International Commission on Stratigraphy. Accessed February 6, 2026. https://stratigraphy.org/timescale/.

Chapter 25: Thaw & Surge

1. Bobrovskiy, Ilya, et al. 2022. "Guts, gut contents, and feeding strategies of Ediacaran animals." Current Biology 32 (24). https://doi.org/10.1016/j.cub.2022.10.051.
2. Bobrovskiy, Ilya, et al. 2018. "Ancient steroids establish the Ediacaran fossil Dickinsonia as one of the earliest animals." Science 361 (6408): 1246–1249. https://doi.org/10.1126/science.aat7228.
3. Clapham, Matthew E., Guy M. Narbonne, and James G. Gehling. 2003. "Paleoecology of the oldest known animal communities: Ediacaran assemblages at Mistaken Point, Newfoundland." Paleobiology 29 (4): 527–544. https://doi.org/10.1666/0094-8373(2003)029<0527:POTOK

A>2.0.CO;2.

4. Cribb, A. T., et al. 2023. "Ediacaran–Cambrian bioturbation did not extensively oxygenate sediments in shallow marine ecosystems." Geobiology 21 (4): 435–453. https://doi.org/10.1111/gbi.12550.
5. Cuthill, Jennifer F., and Simon C. Morris. 2014. "Fractal branching organizations of Ediacaran rangeomorph fronds reveal a lost Proterozoic body plan." Proceedings of the National Academy of Sciences of the United States of America 111 (36): 12883–12888. https://doi.org/10.1073/pnas.1408542111.
6. Darroch, Simon A. F., Marc Laflamme, and Peter J. Wagner. 2018. "High ecological complexity in benthic Ediacaran communities." Nature Ecology & Evolution 2 (10): 1541–1547. https://doi.org/10.1038/s41559-018-0663-7.
7. Darroch, Simon A. F., Emma F. Smith, Marc Laflamme, and Douglas H. Erwin. 2018. "Ediacaran extinction and Cambrian explosion." Trends in Ecology & Evolution 33 (9): 653–663. https://doi.org/10.1016/j.tree.2018.06.003.
8. Darroch, Simon A. F., et al. 2023. "The rangeomorph Pectinifrons abyssalis: Hydrodynamic function at the dawn of animal life." iScience 26 (2): 105989. https://doi.org/10.1016/j.isci.2023.105989.
9. Gehling, James G., Bruce N. Runnegar, and Mary L. Droser. 2014. "Scratch traces of large Ediacara bilaterian animals." Journal of Paleontology 88 (2): 284–298. https://doi.org/10.1666/13-054.
10. Ostrander, Chadlin M., et al. 2023. "Widespread seafloor anoxia during generation of the Ediacaran Shuram carbon isotope excursion." Geobiology 21 (5): 556–570. https://doi.org/10.1111/gbi.12557.
11. Rafferty, John P. 2026. "Cryogenian period." Encyclopaedia Britannica. Accessed February 9, 2026. https://www.britannica.com/science/Cryogenian-Period.
12. Wood, Rachel, et al. 2023. "New Ediacaran biota from the oldest Nama Group, Namibia (Tsaus Mountains), and re-definition of the Nama assemblage." Geological Magazine 160 (9): 1673–1686. https://doi.org/10.1017/S0016756823000638.

13. VAMP (Virtual Australian Museum of Palaeontology), Flinders University. 2026. "Ediacaran fossils." VAMP. Accessed February 9, 2026. https://sites.flinders.edu.au/vamp/fossils/early-life/ediacaran-fauna/.
14. Flinders Ranges Ediacara Foundation. 2026. "Ediacaran fossils." Flinders Ranges Ediacara Foundation. Accessed February 9, 2026. https://ediacarafoundation.org/ediacaran-fossils/.

Chapter 26: Fimbulvetr Aeon

1. Bowles, A. M. C., et al. 2024. "Cryogenian origins of multicellularity in Archaeplastida." Genome Biology and Evolution 16 (2): evae026. https://doi.org/10.1093/gbe/evae026.
2. Choudhary, P., S. Bhatt, and S. Chatterjee. 2024. "From freezing to functioning: Cellular strategies of cold-adapted bacteria for surviving in extreme environments." Archives of Microbiology 206 (7). https://doi.org/10.1007/s00203-024-04058-5.
3. Goddéris, Yves, et al. 2007. "Coupled modeling of global carbon cycle and climate in the Neoproterozoic: Links between Rodinia breakup and major glaciations." Comptes Rendus Geoscience 339. https://doi.org/10.1016/j.crte.2005.12.002.
4. Lund-Hansen, L. C., et al. 2024. "Sea ice as habitat for microalgae, bacteria, virus, fungi, meio- and macrofauna: A review of an extreme environment." Polar Biology. https://doi.org/10.1007/s00300-024-03296-z.
5. Rafferty, John P. 2026. "Cryogenian period." Encyclopaedia Britannica. Accessed February 9, 2026. https://www.britannica.com/science/Cryogenian-Period.
6. Xu, Lingang, et al. 2024. "Termination of Sturtian glaciation with protracted, multiple volcanic eruptions." Earth-Science Reviews 255: 104826. https://doi.org/10.1016/j.earscirev.2024.104826.

7. Yu, Wenchao, et al. 2020. "Cryogenian cap carbonate models: A review and critical assessment." Palaeogeography, Palaeoclimatology, Palaeoecology 552: 109727. https://doi.org/10.1016/j.palaeo.2020.109727.
8. NOAA Office of Ocean Exploration and Research. 2012. "Submarine Ring of Fire 2012: Northeast Lau Basin: Background: Microbiology." NOAA Office of Ocean Exploration and Research. Accessed February 9, 2026. https://oceanexplorer.noaa.gov/explorations/12fire/background/microbio/microbio.html.

Chapter 27: A Gigayear Gestation

1. Alberts, Bruce, et al. 2002. Molecular biology of the cell. 4th ed. New York: Garland Science.
2. Butterfield, Nicholas J. 2000. "Bangiomorpha pubescens n. gen., n. sp.: Implications for the evolution of sex, multicellularity, and the Mesoproterozoic/Neoproterozoic radiation of eukaryotes." Paleobiology 26 (3): 386–404. https://doi.org/10.1666/0094-8373(2000)026<0386:BPNGNS>2.0.CO;2.
3. Lane, Nick. 2011. "Energetics and genetics across the prokaryote–eukaryote divide." Biology Direct 6: 35. https://doi.org/10.1186/1745-6150-6-35.
4. Reinhard, Christopher T., et al. 2013. "Proterozoic ocean redox and biogeochemical stasis." Proceedings of the National Academy of Sciences of the United States of America 110 (14): 5357–5362. https://doi.org/10.1073/pnas.1208622110.
5. Stewart, John H. 2009. Reconstructing Rodinia by fitting Neoproterozoic continental margins. U.S. Geological Survey Open-File Report 2009–1191. https://pubs.usgs.gov/of/2009/1191/.
6. Wei, Wei, et al. 2025. "Dynamic redox evolution in the Middle–Late Mesoproterozoic oceans." Chemical Geology 679: 122685. https://doi.

org/10.1016/j.chemgeo.2025.122685.
7. Xiao, Shuhai, and Qing Tang. 2018. "After the boring billion and before the freezing millions: Evolutionary patterns and innovations in the Tonian Period." Emerging Topics in Life Sciences 2 (2): 161–171. https://doi.org/10.1042/ETLS20170165.
8. Xu, Dongtao, et al. 2023. "Extensive sea-floor oxygenation during the Early Mesoproterozoic." Geochimica et Cosmochimica Acta 354: 186–196. https://doi.org/10.1016/j.gca.2023.06.007.
9. Ye, Yuntao, et al. 2024. "Phosphate oxygen isotopes constrain Mesoproterozoic marine temperatures and the paucity of phosphorite." Chemical Geology 644: 121831. https://doi.org/10.1016/j.chemgeo.2023.121831.

Chapter 28: Atlas Oxidised

1. Allwood, Abigail C., et al. 2006. "Stromatolite reef from the early Archaean era of Australia." Nature 441: 714–718. https://doi.org/10.1038/nature04764.
2. Becker, Richard H., and Robert N. Clayton. 1972. "Carbon isotopic evidence for the origin of a banded iron-formation in Western Australia." Geochimica et Cosmochimica Acta 36 (5): 577–595. https://doi.org/10.1016/0016-7037(72)90077-4.
3. Crowe, Sean A., et al. 2013. "Atmospheric oxygenation three billion years ago." Nature 501 (7468): 535–538. https://doi.org/10.1038/nature12426.
4. Demoulin, Catherine F., et al. 2024. "Oldest thylakoids in fossil cells directly evidence oxygenic photosynthesis." Nature 625: 529–534. https://doi.org/10.1038/s41586-023-06896-7.
5. Harada, Mariko, Eiichi Tajika, and Yasuhito Sekine. 2015. "Transition to an oxygen-rich atmosphere with an extensive overshoot triggered by the Paleoproterozoic snowball Earth." Earth and Planetary Science

Letters 419: 178–186. https://doi.org/10.1016/j.epsl.2015.03.005.

6. Johnstone, Colin P., et al. 2021. "The young Sun's XUV-activity as a constraint for lower CO_2-limits in the Earth's Archean atmosphere." Earth and Planetary Science Letters 576: 117197. https://doi.org/10.1016/j.epsl.2021.117197.
7. Lee, Cin-Ty A., et al. 2016. "Two-step rise of atmospheric oxygen linked to the growth of continents." Nature Geoscience 9: 417–424. https://doi.org/10.1038/ngeo2707.
8. Luo, Genming, et al. 2016. "Rapid oxygenation of Earth's atmosphere 2.33 billion years ago." Science Advances 2 (5): e1600134. https://doi.org/10.1126/sciadv.1600134.
9. Ohmoto, Hiroshi. 1996. "Evidence in pre-2.2 Ga paleosols for the early evolution of atmospheric oxygen and terrestrial biota." Geology 24 (12): 1135–1138. https://doi.org/10.1130/0091-7613(1996)024<1135:EIPGPF>2.3.CO;2.
10. Vitali, Rayanne, et al. 2026. "Combined effects of photorespiration and fire strongly regulate atmospheric oxygen levels." Science Advances 12 (2): eady0542. https://doi.org/10.1126/sciadv.ady0542.
11. StatPearls Publishing. 2022. "Physiology, respiratory rate." In StatPearls. Treasure Island, FL: StatPearls Publishing. https://www.ncbi.nlm.nih.gov/books/NBK537306/.
12. Schmidt-Rohr, Klaus. 2020. "Oxygen is the high-energy molecule powering complex multicellular life: Fundamental corrections to traditional bioenergetics." ACS Omega 5 (5): 2221–2233. https://doi.org/10.1021/acsomega.9b03352.

Chapter 29: Anaerobic Supremacy

1. Allwood, Abigail C., et al. 2006. "Stromatolite reef from the early Archaean era of Australia." Nature 441 (7094): 714–718. https://doi.org/10.1038/nature04764.

2. Bar-On, Yinon M., Rob Phillips, and Ron Milo. 2018. "The biomass distribution on Earth." Proceedings of the National Academy of Sciences of the United States of America 115 (25): 6506–6511. https://doi.org/10.1073/pnas.1711842115.
3. Everts, Sarah. 2016. "Ancient fossils in Greenland suggest evidence for microbial life from 3.7 billion years ago." Chemical & Engineering News 94 (35). https://pubs.acs.org/doi/10.1021/cen-09435-notw3.
4. Everts, Sarah. 2016. "Glycine and phosphorus found on comet." Chemical & Engineering News 94 (22). https://pubs.acs.org/doi/10.1021/cen-09422-scicon003.
5. Konhauser, Kurt O., et al. 2009. "Oceanic nickel depletion and a methanogen famine before the Great Oxidation Event." Nature 458 (7239): 750–753. https://doi.org/10.1038/nature07858.
6. Kvenvolden, Keith A., James G. Lawless, and Cyril Ponnamperuma. 1971. "Nonprotein amino acids in the Murchison meteorite." Proceedings of the National Academy of Sciences of the United States of America 68 (2): 486–490. https://doi.org/10.1073/pnas.68.2.486.
7. Mojzsis, Stephen J., et al. 1996. "Evidence for life on Earth before 3,800 million years ago." Nature 384 (6604): 55–59. https://doi.org/10.1038/384055a0.
8. Sleep, Norman H., et al. 2004. "H_2-rich fluids from serpentinization: Geochemical and biotic implications." Proceedings of the National Academy of Sciences of the United States of America 101 (35): 12818–12823. https://doi.org/10.1073/pnas.0405289101.
9. Thompson, Kevin J., et al. 2019. "Photoferrotrophy, deposition of banded iron formations, and methane production in Archean oceans." Science Advances 5 (2): eaav2869. https://doi.org/10.1126/sciadv.aav2869.
10. Waite, J. Hunter, et al. 2017. "Cassini finds molecular hydrogen in the Enceladus plume: Evidence for hydrothermal processes." Science 356 (6334): 155–159. https://doi.org/10.1126/science.aai8703.

Chapter 30: *Ex Nihilio*

1. Henderson-Sellers, A., and A. W. Schwartz. 1980. "Chemical evolution and ammonia in the early Earth's atmosphere." Nature 287: 526–528. https://doi.org/10.1038/287526a0.
2. Fleury, B., et al. 2017. "Organic chemistry in a CO_2-rich early Earth atmosphere." Earth and Planetary Science Letters 479: 34–42. https://doi.org/10.1016/j.epsl.2017.09.026.
3. Miller, S. L. 1953. "A production of amino acids under possible primitive Earth conditions." Science 117 (3046): 528–529. https://doi.org/10.1126/science.117.3046.528.
4. Rafferty, J. P. n.d. "Hadean eon." Encyclopaedia Britannica. Accessed February 10, 2026. https://www.britannica.com/science/Hadean-Eon.
5. Lissauer, J. J., J. W. Barnes, and J. E. Chambers. 2012. "Obliquity variations of a moonless Earth." Icarus 217 (1): 77–87. https://doi.org/10.1016/j.icarus.2011.10.013.
6. National Aeronautics and Space Administration. n.d. "What is the Late Heavy Bombardment?" NASA Science. Accessed February 10, 2026. https://science.nasa.gov/moon/lunar-craters/what-is-the-late-heavy-bombardment/.
7. Mojzsis, S. J., T. M. Harrison, and R. T. Pidgeon. 2001. "Oxygen-isotope evidence from ancient zircons for liquid water at the Earth's surface 4,300 Myr ago." Nature 409: 178–181. https://doi.org/10.1038/35051557.
8. Abe, S. 2001. "Earth's oldest mineral grains suggest an early start for life." NASA Astrobiology. https://astrobiology.nasa.gov/news/earths-oldest-mineral-grains-suggest-an-early-start-for-life/.
9. Encyclopaedia Britannica. n.d. "Hydrosphere." Encyclopaedia Britannica. Accessed February 10, 2026. https://www.britannica.com/science/hydrosphere.
10. Bell, D. R., and G. R. Rossman. 1992. "Water in Earth's mantle: The role of nominally anhydrous minerals." Science 255 (5050): 1391–1397.

https://doi.org/10.1126/science.255.5050.1391.

11. U.S. Geological Survey. n.d. "Water Q&A: Why is water the 'universal solvent'?" U.S. Geological Survey. Accessed February 10, 2026. https://www.usgs.gov/special-topics/water-science-school/science/water-qa-why-water-universal-solvent.
12. Lane, N. 2017. "Proton gradients at the origin of life." BioEssays 39 (6). https://doi.org/10.1002/bies.201600217.
13. Müller, U. F. 2009. "Evolution of ribozymes in an RNA world." Chemistry & Biology 16 (8): 797–798. https://doi.org/10.1016/j.chembiol.2009.08.001.
14. Schreiber, U. 2025. "The origin of life in the early continental crust: A comprehensive model." Life 15 (3). https://doi.org/10.3390/life15030433.
15. Forterre, P. 2024. "The last universal common ancestor of ribosome-encoding organisms: Portrait of LUCA." Journal of Molecular Evolution. https://doi.org/10.1007/s00239-024-10186-9.
16. Mrnjavac, N., and W. F. Martin. 2025. "GTP before ATP: The energy currency at the origin of genes." Biochimica et Biophysica Acta (BBA) – Bioenergetics 1866 (1). https://doi.org/10.1016/j.bbabio.2024.149514.

Chapter 31: Witnesses to the Eleventh Hour

1. Lindsey, D. A. 2007. "Precambrian time—The story of the early Earth (version 1.0)." U.S. Geological Survey Fact Sheet 2007–3004. https://doi.org/10.3133/fs20073004.
2. Smithsonian National Museum of Natural History. 2017. "Our species arose at least 300,000 years ago." Human Origins Program. June 8, 2017. https://humanorigins.si.edu/research/whats-hot-human-origins/our-species-arose-least-300000-years-ago.
3. Hublin, Jean-Jacques, et al. 2017. "New fossils from Jebel Irhoud, Morocco and the pan-African origin of Homo sapiens." Nature 546

(7657): 289–292. https://doi.org/10.1038/nature22336.
4. Richter, Daniel, et al. 2017. "The age of the hominin fossils from Jebel Irhoud, Morocco, and the origins of the Middle Stone Age." Nature 546 (7657): 293–296. https://doi.org/10.1038/nature22335.
5. The Holy Bible. n.d. "Matthew 20:9." BibleGateway. Accessed February 10, 2026. https://www.biblegateway.com/verse/en/Matthew%2020%3A9.
6. Murphy, J. B. n.d. "Plate tectonics." Encyclopaedia Britannica. Accessed February 10, 2026. https://www.britannica.com/science/plate-tectonics.
7. U.S. Geological Survey. n.d. "Mud fossils—Background." U.S. Geological Survey. Accessed February 10, 2026. https://www.usgs.gov/educational-resources/mud-fossils-background.
8. University of California Museum of Paleontology, University of California, Berkeley. n.d. "The origin of birds." Understanding Evolution. Accessed February 10, 2026. https://evolution.berkeley.edu/what-are-evograms/the-origin-of-birds/.
9. U.S. Geological Survey. 2008. "Limestone." U.S. Geological Survey Fact Sheet 2008–3089. https://pubs.usgs.gov/fs/2008/3089/fs2008-3089.pdf.

Chapter 32: The Ledger of Prometheus

1. Curie, Marie. 1911. "Radium and the new concepts in chemistry." Nobel Prize Lecture. December 11, 1911. https://www.nobelprize.org/prizes/chemistry/1911/marie-curie/lecture/.
2. Encyclopaedia Britannica Editors. n.d. "Marie Curie." Encyclopaedia Britannica. Accessed February 10, 2026. https://www.britannica.com/biography/Marie-Curie.
3. Wei, Wei, et al. 2026. "Ionizing radiation: Molecular mechanisms, biological effects, and therapeutic targets." Molecular Biomedicine 7

(3). https://doi.org/10.1186/s43556-025-00358-4.

4. Williams, Henry Smith, and Edward Huntington Williams. 1999. A history of science. Vol. 4, Modern development of the chemical and biological sciences. Project Gutenberg. https://www.gutenberg.org/ebooks/1708.
5. Noble, Caitlin A., et al. 2025. "The history of the microscope reflects advances in science and medicine." Seminars in Diagnostic Pathology 42 (2): 150–153. https://doi.org/10.1053/j.semdp.2024.01.002.
6. Joo, Jeong H., Soogene Lee, and Keun P. Kim. 2025. "Precision gene editing: The power of CRISPR-Cas in modern genetics." Molecular Therapy – Nucleic Acids 27: 102733. https://doi.org/10.1016/j.omtn.2025.102733.
7. Bach, Benedikt G., et al. 2023. "Visualizing quantum circuit probability: Estimating quantum state complexity for quantum program synthesis." Entropy 25 (5). https://doi.org/10.3390/e25050763.
8. Kong, Xin, et al. 2023. "Advances of medical nanorobots for future cancer treatments." Journal of Hematology & Oncology 16. https://doi.org/10.1186/s13045-023-01463-z.
9. U.S. Government Accountability Office. 2023. Fusion energy: Potentially transformative technology still faces fundamental challenges. GAO-23-105813. https://www.gao.gov/products/gao-23-105813.
10. Bednarz, Eliza, et al. 2025. "Marine cloud brightening may cool the Earth—but could impact the ozone layer." NOAA Chemical Sciences Laboratory. May 14, 2025. https://csl.noaa.gov/news/2025/428_0514.html.
11. Cervera, Rafael, et al. 2024. "The artificial tree: Integrating microalgae into sustainable architecture for CO_2 capture and urban efficiency—A comprehensive analysis." Buildings 14 (12). https://doi.org/10.3390/buildings14124045.
12. Kansara, Ketan, et al. 2025. "DNA nano-biomaterials based futuristic technologies for tissue engineering and regenerative therapeutics." Small 21 (31). https://doi.org/10.1002/smll.202504361.
13. Agudelo-Toro, Andres, et al. 2024. "Accurate neural control of a hand

prosthesis by posture-related activity in the primate grasping circuit." Neuron 112 (24): 4115–4129.e8. https://doi.org/10.1016/j.neuron.2024.09.018.

14. Global Hunger Index. 2024. "Global Hunger Index." Global Hunger Index. Accessed February 10, 2026. https://www.globalhungerindex.org/.

Chapter 33: Mythopoeia of the Primate

1. Bednarik, Robert G. 2015. "Fire in the mind: Changing understandings of fire in Western civilization." Medical Humanities 41 (2): 115–120. https://doi.org/10.1136/medhum-2014-010635.
2. Fuentes, Agustín. n.d. "How did belief evolve?" SAPIENS. Accessed February 10, 2026. https://www.sapiens.org/biology/religion-origins/.
3. Reuell, Peter. 2016. "Religion as social unifier." Harvard Gazette, July 29, 2016. https://news.harvard.edu/gazette/story/2016/07/religion-as-social-unifier/.
4. Laidlaw, James. 2017. "Ethics/morality." In The Open Encyclopedia of Anthropology. https://www.anthroencyclopedia.com/entry/ethics-morality.
5. Metropolitan Museum of Art. n.d. "The Akwanshi stone monoliths of the Cross River region in Nigeria and Cameroon." The Met. Accessed February 10, 2026. https://www.metmuseum.org/essays/akwanshi-stone-monoliths.
6. Pettitt, Paul, and Paul Bahn. 2019. "Homo neanderthalensis and the evolutionary origins of ritual in Homo sapiens." Frontiers in Psychology 10: 2673. https://doi.org/10.3389/fpsyg.2019.02673.
7. Whitehouse, Harvey. n.d. "Ritual modes: Divergent modes of ritual, social cohesion, prosociality, and conflict." Institute of Social and Cultural Anthropology, University of Oxford. https://www.anthro.ox.ac.uk/ritual-modes-divergent-modes-of-ritual-social-cohesion-pros

ociality-and-conflict.

8. Whitehouse, Harvey. 2022. "The role of ritual in the evolution of social complexity." In The Oxford handbook of the archaeology of ritual and religion, edited by Courtney Nimura, 395–412. Oxford: Oxford University Press.
9. Encyclopaedia Britannica Editors. n.d. "Divine right of kings." Encyclopaedia Britannica. Accessed February 10, 2026. https://www.britannica.com/topic/divine-right-of-kings.

Chapter 34: A Requiem for the Pantheon

1. Oxford University Museum of Natural History. n.d. "The Great Debate." Accessed February 10, 2026. https://oumnh.ox.ac.uk/great-debate.
2. Chamberlain, Lesley. 2012. "The political message of Nietzsche's 'God is dead'." The Guardian, February 7, 2012. https://www.theguardian.com/commentisfree/belief/2012/feb/07/political-message-nietzsche-god-is-dead.
3. Ebel, Jonathan H. 2010. Faith in the fight: Religion and the American soldier in the Great War. Princeton, NJ: Princeton University Press.
4. Higgins, E. Tory. 2019. "The roots of human shared reality." In Shared reality: What makes us strong and tears us apart. Oxford: Oxford University Press. https://academic.oup.com/book/35191/chapter/299567693.
5. Intergovernmental Science-Policy Platform on Biodiversity and Ecosystem Services (IPBES). 2019. Global assessment report on biodiversity and ecosystem services: Summary for policymakers. https://files.ipbes.net/ipbes-web-prod-public-files/inline/files/ipbes_global_assessment_report_summary_for_policymakers.pdf.
6. Mind and Life Education Research Network. 2012. "Contemplative practices and mental training: Prospects for American education." Child Development Perspectives 6 (2): 146–153. https://doi.org/10.

1111/j.1750-8606.2012.00240.x.
7. Taylor, Paul D. 2004. "Extinction and the fossil record." In Extinctions in the history of life, edited by Paul D. Taylor. Cambridge: Cambridge University Press. https://www.cambridge.org/core/books/extinctions-in-the-history-of-life/extinction-and-the-fossil-record/6F1E14342C6A1FEE1EB86DB0E70D58A2.
8. von Baeyer, Hans Christian. 1998. "A century of physics." APS News (American Physical Society). https://www.aps.org/publications/apsnews/199803/prologue.cfm.
9. Voegelin, Eric. 2000. "The political religions." In Modernity without restraint: The political religions; The new science of politics; and Science, politics and Gnosticism, edited by Manfred Henningsen. Columbia: University of Missouri Press.
10. Gentile, Emilio. 2006. Politics as religion. Princeton, NJ: Princeton University Press.
11. Esposito, John L., ed. 1983. Voices of resurgent Islam. New York: Oxford University Press.

Chapter 35: Simulation and Stratum

1. Bergström, Anders, et al. 2022. "Grey wolf genomic history reveals a dual ancestry of dogs." Nature 607 (7918): 313–320. https://doi.org/10.1038/s41586-022-04824-9.
2. Bostrom, Nick. 2003. "Are we living in a computer simulation?" The Philosophical Quarterly 53 (211): 243–255. https://doi.org/10.1111/1467-9213.00309.
3. Encyclopaedia Britannica. n.d. "Myth of the cave." Accessed February 10, 2026. https://www.britannica.com/topic/myth-of-the-cave.
4. Faizal, Mir, et al. 2025. "Consequences of undecidability in physics on the theory of everything." arXiv. https://arxiv.org/abs/2507.22950.
5. Kaufmann, Angelica, et al. 2025. "What holds groups together? How

interdependence shapes group-living." Behavioral and Brain Sciences 48: e172. https://doi.org/10.1017/S0140525X25101775.
6. Öhman, Arne, et al. 2007. "On the unconscious subcortical origin of human fear." Physiology & Behavior 92 (1–2): 180–185. https://doi.org/10.1016/j.physbeh.2007.05.057.
7. Wang, Fan, and Shaoshan Liu. 2025. "Environmental embodied AI." Communications of the ACM (blog), April 3, 2025. https://cacm.acm.org/blogcacm/environmental-embodied-ai/.
8. Wheeler, John Archibald. 1990. "Information, physics, quantum: The search for links." In Complexity, entropy, and the physics of information, edited by Wojciech H. Zurek, 3–28. Redwood City, CA: Addison-Wesley.

Chapter 36: Through the Celestial Looking Glass

1. Crawford, Ian A., and Dirk Schulze-Makuch. 2024. "Is the apparent absence of extraterrestrial technological civilizations down to the zoo hypothesis or nothing?" Nature Astronomy 8: 44–49. https://doi.org/10.1038/s41550-023-02134-2.
2. Des Marais, David J. 1991. "Microbial mats, stromatolites and the rise of oxygen in the Precambrian atmosphere." Global and Planetary Change 5: 93–96. https://doi.org/10.1016/0921-8181(91)90130-O.
3. Encyclopaedia Britannica. 2026. "Fermi paradox." Accessed February 10, 2026. https://www.britannica.com/science/Fermi-paradox.
4. Martin, Andrew, and Andrew McMinn. 2018. "Sea ice, extremophiles and life on extra-terrestrial ocean worlds." International Journal of Astrobiology 17 (1): 1–16. https://doi.org/10.1017/S1473550416000483.
5. National Aeronautics and Space Administration. 2015. "Apollo 11 mission overview." Accessed February 10, 2026. https://www.nasa.gov/history/apollo-11-mission-overview/.
6. National Aeronautics and Space Administration. n.d. "Exoplanet

discovery statistics." Accessed February 10, 2026. https://science.nasa.gov/exoplanets/.

7. National Aeronautics and Space Administration. n.d. "How big is the Milky Way?" Accessed February 10, 2026. https://science.nasa.gov/universe/exoplanets/our-milky-way-galaxy-how-big-is-space/.
8. National Aeronautics and Space Administration Jet Propulsion Laboratory. 2012. "Study shows our galaxy has at least 100 billion planets." Accessed February 10, 2026. https://www.jpl.nasa.gov/news/study-shows-our-galaxy-has-at-least-100-billion-planets/.
9. National Geographic Society. 2023. "Formation of Earth." Accessed February 10, 2026. https://education.nationalgeographic.org/resource/formation-earth/10th-grade/.
10. SETI Institute. n.d. "Drake equation." Accessed February 10, 2026. https://www.seti.org/research/seti-101/drake-equation/.

Chapter 37: The Bones Above Us

1. Dalrymple, G. Brent. 1999. "The age of the Earth—a summary." The Paleontological Society Papers 5: 1–22. https://doi.org/10.1017/S1089332600000516.
2. DeBenedictis, Erika Alden, et al. 2025. "The case for Mars terraforming research." Nature Astronomy 9. https://doi.org/10.1038/s41550-025-02548-0.
3. Encyclopaedia Britannica. 2026. "List of dynasties of ancient Egypt." Accessed February 10, 2026. https://www.britannica.com/topic/list-of-dynasties-of-ancient-Egypt.
4. Encyclopaedia Britannica. n.d. "Sun: Evolution." Accessed February 10, 2026. https://www.britannica.com/place/Sun/Evolution.
5. Environmental Protection Agency. 2025. "Overview of greenhouse gases." Accessed February 10, 2026. https://www.epa.gov/ghgemissions/overview-greenhouse-gases.

6. Hublin, Jean-Jacques, et al. 2017. "New fossils from Jebel Irhoud, Morocco and the Pan-African origin of Homo sapiens." Nature 546 (7657): 289–292. https://doi.org/10.1038/nature22336.
7. James, David H., et al. 2025. "Classic Maya response to multiyear seasonal droughts in northwest Yucatán, Mexico." Science Advances. https://doi.org/10.1126/sciadv.adw7661.
8. Janik, Vincent M. 2014. "Cetacean vocal learning and communication." Current Opinion in Neurobiology 28: 60–65. https://doi.org/10.1016/j.conb.2014.06.010.
9. Lemoine, Rhys Taylor, Robert Buitenwerf, and Jens-Christian Svenning. 2023. "Megafauna extinctions in the late-Quaternary are linked to human range expansion, not climate change." The Anthropocene 44: 100403. https://doi.org/10.1016/j.ancene.2023.100403.
10. National Park Service. 2025. "Mass extinctions through geologic time." Accessed February 10, 2026. https://www.nps.gov/subjects/fossils/mass-extinctions-through-geologic-time.htm.
11. Pogge, Richard W. 2007. "The once & future Sun." Accessed February 10, 2026. https://www.astronomy.ohio-state.edu/pogge.1/Lectures/vistas97.html.
12. Prieur, Jacques, and Simone Pika. 2020. "Gorillas' (Gorilla g. gorilla) knowledge of conspecifics' affordances: Intraspecific social tool use for food acquisition." Primates 61 (4): 583–591. https://doi.org/10.1007/s10329-020-00805-6.
13. Simonetti, Cristián, and Matt Edgeworth. 2022. "Concrete: A stratigraphic marker for the Anthropocene." Anthropocene Curriculum. Accessed February 10, 2026. https://www.anthropocene-curriculum.org/contribution/concrete-a-stratigraphic-marker-for-the-anthropocene.
14. Smithsonian Institution, Human Origins Program. 2017. "Our species arose at least 300,000 years ago." Accessed February 10, 2026. https://humanorigins.si.edu/research/whats-hot-human-origins/our-species-arose-least-300000-years-ago.
15. Wascher, Claudia A. F., et al. 2025. "Vocal mimicry in corvids." Animal

Cognition. https://doi.org/10.1007/s10071-025-02006-4.

Chapter 38: *Aetas Fragilis Hominum*

1. Bejan, Adrian. 2020. "Human evolution is biological & technological evolution." BioSystems 195: 104156. https://doi.org/10.1016/j.biosystems.2020.104156.
2. Bellaby, Ross. 2024. "The ethical problems of 'intelligence–AI'." International Affairs 100 (6): 2525–2542. https://doi.org/10.1093/ia/iiae227.
3. Galor, Oded. 2011. Unified growth theory. Princeton, NJ: Princeton University Press.
4. Encyclopaedia Britannica. 2026. "Primate: Classification." Accessed February 10, 2026. https://www.britannica.com/animal/primate-mammal/Classification.
5. Kelsey, Jason. n.d. "Biodiversity, evolution, and extinction." In Environmental Science: An Evidence-Based Study of Earth's Natural Systems. Accessed February 10, 2026. https://open.muhlenberg.pub/environmentalscience/.
6. Lee, Michael S. Y., Julien Soubrier, and Gregory D. Edgecombe. 2013. "Rates of phenotypic and genomic evolution during the Cambrian explosion." Current Biology 23 (19): 1889–1895. https://doi.org/10.1016/j.cub.2013.07.055.
7. Liao, Hongyan, Christopher J. Lyon, Binwu Ying, and Tony Hu. 2024. "Climate change, its impact on emerging infectious diseases and new technologies to combat the challenge." Emerging Microbes & Infections 13 (1): 2356143. https://doi.org/10.1080/22221751.2024.2356143.
8. Organisation for Economic Co-operation and Development. 2024. "Declining fertility rates put prosperity of future generations at risk." Accessed February 10, 2026. https://www.oecd.org/en/about/news/pre

ss-releases/2024/06/declining-fertility-rates-put-prosperity-of-future-generations-at-risk.html.

9. ÓhÉigeartaigh, Sean. 2025. "Extinction of the human species: What could cause it and how likely is it to occur?" Cambridge Prisms: Extinction 3: e4. https://doi.org/10.1017/ext.2025.4.
10. Rampino, Michael R., Ken Caldeira, and Andreas Prokoph. 2019. "What causes mass extinctions? Large asteroid/comet impacts, flood-basalt volcanism, and ocean anoxia—correlations and cycles." In 250 Million Years of Earth History in Central Italy: Celebrating 25 Years of the Geological Observatory of Coldigioco, edited by Christian Koeberl and David M. Bice, 271–302. Geological Society of America Special Paper 542. https://doi.org/10.1130/2019.2542(14).
11. Samuel, Ugochukwu Daniel, Omobolanle Comfort Adegbola, Iyabo Mariam Shittu, Bankole John Falana, and Favour Esther Obidiah. 2024. "Economic growth and environmental sustainability: Empirical evidence from selected African countries." Path of Science 10. https://doi.org/10.31018/ps.2024.3040.
12. Smart, Paul R. 2018. "Human-extended machine cognition." Cognitive Systems Research 49: 9–23. https://doi.org/10.1016/j.cogsys.2017.11.001.
13. Smith, Daniel, et al. 2017. "Cooperation and the evolution of hunter-gatherer storytelling." Nature Communications 8: 1853. https://doi.org/10.1038/s41467-017-02036-8.
14. Vitousek, Peter M., Harold A. Mooney, Jane Lubchenco, and Jerry M. Melillo. 1997. "Human domination of Earth's ecosystems." Science 277 (5325): 494–499. https://doi.org/10.1126/science.277.5325.494.

Further Reading

Human History and Civilisation

Harari, Yuval Noah. Sapiens: A Brief History of Humankind.

Harari, Yuval Noah. Homo Deus: A Brief History of Tomorrow.

Holland, Tom. Dominion: The Making of the Western Mind.

Ryan, Christopher. Civilised to Death: The Price of Progress.

Ryan, Christopher, and Cacilda Jethá. Sex at Dawn: The Prehistoric Origins of Modern Sexuality.

Pinker, Steven. Enlightenment Now: The Case for Reason, Science, Humanism, and Progress.

Evolution and the Deep Past

Dawkins, Richard. The Selfish Gene.

Dawkins, Richard. River Out of Eden: A Darwinian View of Life.

Dawkins, Richard. The Ancestor's Tale: A Pilgrimage to the Dawn of Evolution.

Lane, Nick. The Vital Question: Why Is Life the Way It Is?

Sagan, Carl. The Dragons of Eden: Speculations on the Evolution of Human Intelligence.

Howard, Jules. Death on Earth: Adventures in Evolution and Mortality.

Panciroli, Elsa. The Beasts Before Us: The Untold Story of Mammal Origins and Evolution.

MacPhee, R. D. E. The End of the Megafauna: The Fate of the World's Hugest, Fiercest, and Strangest Animals.

Cregan-Reid, Vybarr. Primate Change: How the World We Made Is Remaking Us.

Brusatte, Steve. The Rise and Reign of the Mammals: A New History, from the Shadow of the Dinosaurs to Us.

Brusatte, Steve. The Dinosaurs Rediscovered: The Scientific Revolution in Palaeontology.

Shapiro, Beth. How to Clone a Mammoth: The Science of De-Extinction.

Randall, Lisa. Dark Matter and the Dinosaurs: The Astounding Interconnectedness of the Universe.

Ecology, Extinction, and the Living World

Kolbert, Elizabeth. The Sixth Extinction: An Unnatural History.

Jepson, Paul, and Cain Blythe. Rewilding: The Radical New Science of Ecological Recovery.

Dunn, Rob. A Natural History of the Future: What the Laws of Biology Tell

Us about the Destiny of the Human Species.

Sheldrake, Merlin. Entangled Life: How Fungi Make Our Worlds, Change Our Minds and Shape Our Futures.

Consciousness, Technology, and the Future

Hanson, Robin. The Age of Em: Work, Love, and Life When Robots Rule the Earth.

Tegmark, Max. Life 3.0: Being Human in the Age of Artificial Intelligence.

De Waal, Frans. Our Inner Ape: A Leading Primatologist Explains Why We Are Who We Are.

Baggini, Julian. How the World Thinks: A Global History of Philosophy.

Glossary

Abiogenesis

The process by which life arises from non-living chemical systems. Abiogenesis describes the transition from prebiotic chemistry to the first self-sustaining, evolving biological entities on the early Earth.

Anaerobic

Describing organisms, environments, or biological processes that operate without oxygen. Many early life forms were anaerobic, reflecting conditions on the young Earth before oxygen accumulated in the atmosphere.

Anthropocene

A proposed geological epoch defined by the global influence of human activity on Earth's climate, ecosystems, and geological systems. The Anthropocene reflects the long-term planetary imprint of industrialisation, land use change, and large-scale resource extraction.

Anoxia / anoxic

The complete absence of oxygen within a given environment. Anoxic conditions occur in certain marine basins, sediments, and deep waters, shaping which organisms can survive and how chemical cycles operate.

Archosaurs

A major group of reptiles that includes crocodilians, birds, and all non-avian dinosaurs. Archosaurs rose to ecological dominance after the Permian mass extinction and shaped terrestrial ecosystems throughout the Mesozoic Era.

Banded iron formations (BIFs)

Layered sedimentary rocks composed of alternating iron-rich and silica-rich bands. Banded iron formations record early interactions between microbial oxygen production and dissolved iron in the ancient oceans.

Biodiversity

The variety of life across genes, species, and ecosystems within a given region or across the planet. Biodiversity reflects both evolutionary history and the ecological conditions that sustain living systems.

Biogeochemical cycle

The movement of chemical elements and compounds through living organisms, the atmosphere, the oceans, and the solid Earth. Biogeochemical cycles regulate the availability of nutrients and shape the stability of planetary environments.

Biomineralization

The biological production of mineral structures such as shells, skeletons, and teeth. Biomineralization alters the chemistry of oceans and sediments while reshaping ecological interactions through the emergence of hard body parts.

Biosphere

The global sum of all ecosystems and living organisms on Earth. The biosphere interacts continuously with the atmosphere, hydrosphere, and lithosphere through energy flow and chemical exchange.

Cambrian explosion

A geologically brief interval around 541 million years ago marked by the rapid appearance of most major animal body plans in the fossil record. The Cambrian explosion reflects a profound reorganisation of marine ecosystems and evolutionary dynamics.

Carbon cycle

The circulation of carbon between the atmosphere, oceans, living organisms, soils, and rocks. The carbon cycle regulates climate, supports biological productivity, and links short-term ecological processes with long-term geological change.

Carboniferous

A geological period spanning roughly 359 to 299 million years ago, characterised by widespread tropical forests, high atmospheric oxygen levels, and the formation of many of today's coal deposits.

Cryogenian

A Neoproterozoic period marked by repeated global glaciations, including episodes when ice extended to low latitudes. The Cryogenian shaped ocean chemistry, climate dynamics, and the evolutionary context of early complex life.

Deoxygenation

The decline of oxygen levels in marine or atmospheric environments.

Deoxygenation alters habitat structure, restricts aerobic life, and disrupts ecological and chemical equilibria in aquatic systems.

DNA

Deoxyribonucleic acid, the molecular carrier of genetic information in all known life. DNA stores, replicates, and transmits the instructions that guide cellular structure, function, and inheritance.

Ediacaran biota

Assemblages of large, soft-bodied organisms that lived before the Cambrian Period. The Ediacaran biota represent early experiments in multicellular life and ecological organisation.

End-Permian extinction

The most severe mass extinction in Earth's history, occurring around 252 million years ago. The End-Permian extinction eliminated the majority of marine species and profoundly reshaped terrestrial ecosystems.

Extinction event

A period of rapid and widespread species loss across multiple environments. Extinction events restructure ecosystems, redirect evolutionary pathways, and alter the long-term composition of life on Earth.

Extremophiles

Organisms adapted to thrive in environments characterised by extreme heat, salinity, pressure, acidity, or radiation. Extremophiles reveal the breadth of physiological possibilities within life and inform models of habitability beyond Earth.

Feedback loop (climate)

A process in which an initial change in the climate system triggers effects that further influence that change. Feedback loops shape climate sensitivity by amplifying or damping shifts in temperature, ice cover, atmospheric composition, and ocean circulation.

Fossil record

The total body of preserved remains, traces, and impressions of past life found within geological strata. The fossil record provides the primary physical archive for reconstructing the history of life and patterns of extinction and diversification.

Geologic time scale

The chronological framework used to divide Earth's history into eons, eras, periods, epochs, and ages. The geologic time scale organises deep time according to stratigraphy, radiometric dating, and major transitions in Earth's physical and biological systems.

Greenhouse gases

Atmospheric gases that absorb and re-emit infrared radiation, thereby influencing the retention of heat within Earth's climate system. Greenhouse gases regulate planetary temperature through their effects on radiative balance.

Hominin

Any species more closely related to modern humans than to chimpanzees. Hominins include extinct human relatives and ancestors that display varying combinations of bipedal locomotion, brain expansion, and tool-associated behaviours.

Hydrothermal vents

Seafloor systems where geothermally heated fluids emerge from the oceanic crust. Hydrothermal vents support chemosynthetic ecosystems and provide chemically rich environments relevant to models of early metabolic evolution.

Isotope

Atoms of the same element that share the same number of protons but differ in neutron number. Isotopic ratios preserve information about geological processes, environmental conditions, and biological activity across deep time.

Mass extinction

A geologically brief interval marked by the loss of a large proportion of species across multiple habitats and taxonomic groups. Mass extinctions restructure ecosystems and reset evolutionary trajectories on a planetary scale.

Megafauna

Large-bodied animal species within a given ecosystem or geological interval. Megafauna shape landscapes, nutrient flows, and food-web structure through their ecological roles and population dynamics.

Metabolism

The network of chemical reactions that sustains life within cells and organisms. Metabolism governs energy transformation, material exchange, and the maintenance of biological structure over time.

Neoproterozoic

The final era of the Proterozoic Eon, spanning roughly 1,000 to 541 million years ago. The Neoproterozoic encompasses major climatic upheavals, large-scale glaciations, and the earliest widespread appearance of complex multicellular life.

Ocean anoxia

A state in which seawater contains little to no dissolved oxygen. Ocean anoxia alters marine chemistry, constrains animal life, and has repeatedly reshaped marine ecosystems across Earth's history.

Oxidation

A chemical process involving the loss of electrons during a reaction. Oxidation governs energy transfer in biological systems and drives major geochemical transformations within Earth's atmosphere, oceans, and crust.

Photosynthesis

A biological process by which certain organisms convert light energy into chemical energy stored in organic molecules. Photosynthesis has shaped Earth's surface environment by supplying primary biological energy and altering atmospheric composition.

Plate tectonics

The large-scale movement of Earth's lithospheric plates across the planet's surface. Plate tectonics governs the distribution of continents and oceans, regulates long-term carbon cycling, and influences patterns of volcanism, mountain building, and basin formation.

Proterozoic

The eon of Earth history extending from about 2.5 billion to 541 million years ago. The Proterozoic records the stabilisation of continental crust, major shifts in ocean and atmospheric chemistry, and the gradual emergence of complex life.

Radiometric dating

A method of determining the age of rocks and minerals through the decay of radioactive isotopes. Radiometric dating anchors the geologic time scale by providing absolute ages for key stratigraphic boundaries and evolutionary events.

Redox (reduction–oxidation)

Chemical reactions involving the transfer of electrons between substances. Redox processes structure metabolic pathways and regulate the chemical state of Earth's surface environments.

Stratigraphy

The study of rock layers and their spatial and temporal relationships. Stratigraphy reconstructs sequences of deposition, erosion, and environmental change across geological time.

Supercontinent

A landmass formed by the assembly of most or all of Earth's continental crust into a single configuration. Supercontinents reorganise climate patterns, ocean circulation, and biological dispersal over deep time.

Terraforming

The theoretical modification of a planetary environment to support Earth-like surface conditions. Terraforming concepts draw on planetary science, atmospheric chemistry, and ecological modelling to explore the long-term habitability of other worlds.

Trophic level

The position an organism occupies within a food web based on how it acquires energy. Trophic levels structure energy flow and ecological relationships within ecosystems.

Universal common ancestor (LUCA)

The ancestral population from which all extant cellular life descends. LUCA represents the earliest known point of shared biochemical inheritance across the living world.

About the Author

George Wakeman is the creator of *Extinct Animals Facts*, a science communication project he has run since 2020. His work examines prehistory, extinction, and deep time, with a focus on how physical evidence, evolutionary processes, and planetary change shape the living world.

He studied geography with an emphasis on Earth sciences, deep time, and human responses to climate change, completing his undergraduate dissertation on Holocene fluvial deposits in the River Mersey and their relationship to the emergence of early agriculture. His academic path later shifted toward urban planning, where he completed a master's degree and worked professionally within modern systems of land use and development.

That experience sharpened his interest in the long arc of human presence within Earth's history. He left the field to pursue independent research and writing full-time, grounding his work in primary scientific literature while building *Extinct Animals Facts* into a sustained public archive of prehistoric life. Questions from readers repeatedly drove deeper investigation, turning communication into a process of ongoing inquiry.

Written over two years, *The Bones Beneath Us* situates extinction, evolution, and humanity within a continuous narrative of deep time, shaped by material evidence, scientific uncertainty, and the contingent history of life on Earth.

He publishes articles and a newsletter at www.*extinctanimalsfacts.com*.

Instagram: *@extinctanimalsfacts*

www.ingramcontent.com/pod-product-compliance
Ingram Content Group UK Ltd.
Pitfield, Milton Keynes, MK11 3LW, UK
UKHW040233290926
15471UKWH00002B/169

9 781919 358628